AQA Mathematics

for GCSE

Exclusively endorsed and approved by AQA

Homework Book

D1796103

Series Editor
Paul Metcalf

Series Advisor
David Hodgson

Lead Author
Margaret Thornton

June Haighton
Anne Haworth
Janice Johns
Steven Lomax
Andrew Manning
Kathryn Scott
Chris Sherrington
Mark Willis

FOUNDATION
Linear 2

Nelson Thornes
a Wolters Kluwer business

Published in 2006 by:
Nelson Thornes Ltd
Delta Place   .
27 Bath Road
CHELTENHAM
GL53 7TH
United Kingdom

06 07 08 09 10 / 10 9 8 7 6 5 4 3 2 1

A catalogue record for this book is available from the British Library.

ISBN 0 7487 9780 7

Cover photograph: Seals by Stephen Frink/Digital Vision LU (NT)
Page make-up by MCS Publishing Services Ltd, Salisbury, Wiltshire

Printed and bound in Spain by GraphyCems

Acknowledgements

The authors and publishers wish to thank the following for their contribution:
David Bowles for providing the Assess questions
David Hodgson for reviewing draft manuscripts

Thank you to the following schools:
Little Heath School, Reading
The Kingswinford School, Dudley
Thorne Grammar School, Doncaster

The publishers have made every effort to contact copyright holders but apologise if any have been overlooked.

Contents

Introduction

This book contains homework that allows you to practise what you have just learned. Each chapter is divided into sections that correspond to the numbered Learn topics for the matching chapter in the Students' Book.

 Means that these questions should be attempted with a calculator.

 Means that these questions are practice for the non-calculator paper in the exam and should be attempted without a calculator.

1 ◄——— Underlined questions are harder questions.

Coursework

This section explains the coursework mark scheme and features three Using and applying mathematics mini-coursework tasks.

Homework 1

1 The following solid has a volume of 6 cm^3.

Draw three more solids with volumes of 6 cm^3.

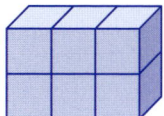

2 Calculate the volume of this solid.

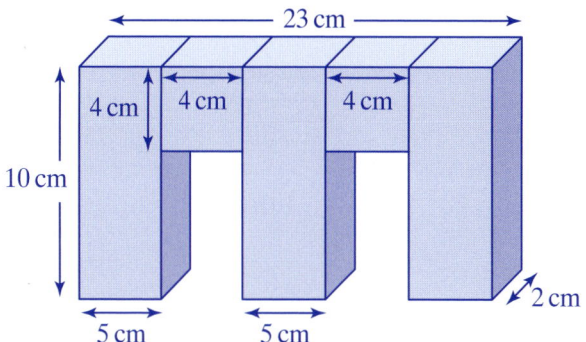

3 An Olympic swimming pool has dimensions 50 m × 25 m × 2 m.

a Calculate the capacity of the swimming pool in litres.

b The pool is filled at a rate of 2000 litres/min.
How many hours does it take to fill the pool?

4 Copy and complete the table:

	Solid	Length	Width	Height	Volume
a	Cuboid	4 cm	2 cm	5 cm	
b	Cuboid	5 cm		2 cm	40 cm^3
c	Cuboid	10 mm	4 mm		160 mm^3
d	Cube	1 cm			
e	Cuboid	2 m		2 m	18 m^3
f	Cuboid	0.25 cm	2 cm		6 cm^3
g	Cube				1 litre

Homework 2

1 Calculate the volume of each of the following solids.

a

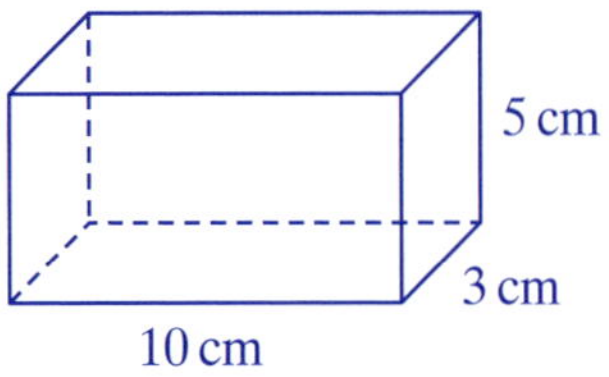

5 cm

3 cm

10 cm

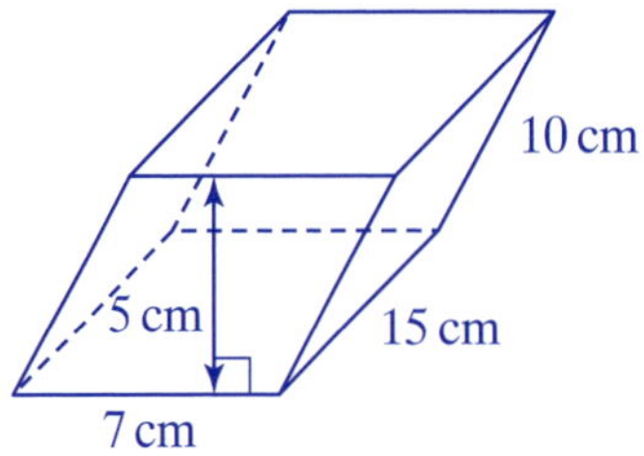

d

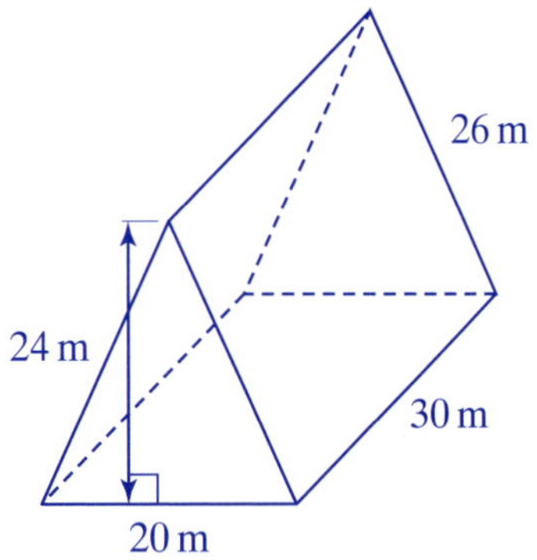

26 m

24 m

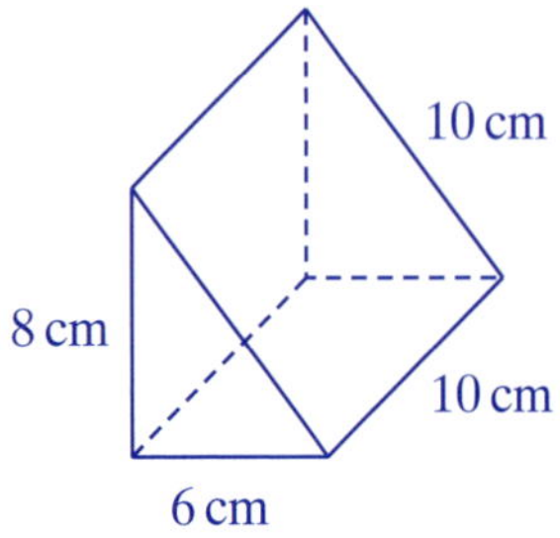

30 m

20 m

b

10 cm

8 cm

10 cm

6 cm

e

4 m

5 m

4 m

15 m

10 m

c

10 cm

5 cm

15 cm

7 cm

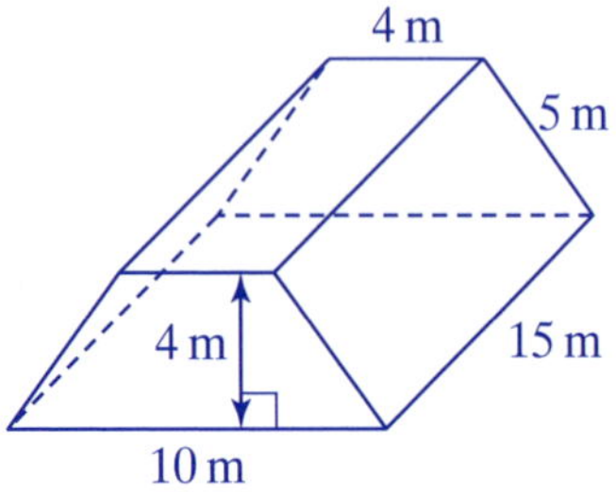

2 Find three different solids with a volume of 120 cm^3.

3 Beth is having a paddling pool party. The party starts in 30 minutes. Beth's dad is filling the paddling pool with his hosepipe at a rate of 150 litres/min. Will it be ready in time?

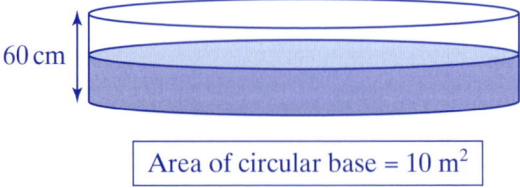

60 cm

Area of circular base = 10 m^2

4 Find the volume of the following cylinders.

Leave your answers in terms of π

 a   6 cm

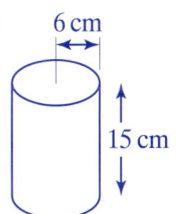   15 cm

b

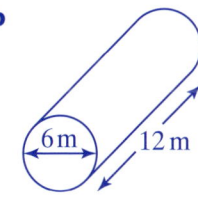   6 m   12 m

5 Copy and complete the table:

Solid	Base of cross section	Perpendicular height of cross section	Area of cross section	Length	Volume
	4 cm	2 cm			24 cm³
	10 cm	5 cm		10 cm	
	5 cm			8 cm	80 cm³

Homework 3

 1   **a** Calculate the surface area of each of the following prisms, stating the units in your answer:

i

    4 cm
    3 cm
8 cm

ii

   8 m
    2 m
    2 m

iii

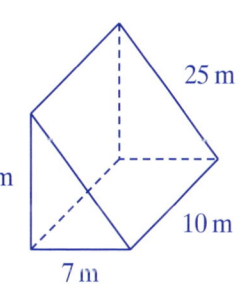   25 m
24 m   10 m
    7 m

b Convert your answer to part **a ii** into cm².

2 These five special solids, with identical regular polygons as faces, are named after the Greek mathematician Plato – they are known as **Platonic solids**. In the diagrams below, each face has an area of 20 cm^2.

  a Calculate the surface area of each solid.

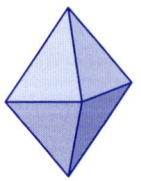 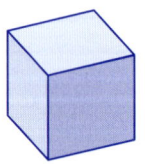  

| Tetrahedron (4 triangles) | Octahedron (8 triangles) | Cube (6 squares) | Icosahedron (20 triangles) | Dodecahedron (12 pentagons) |

  b Use the internet to find a net for each of these Platonic solids.

For questions 3 to 7, match the solid with a correct name, a correct surface area and a correct volume.
You must remember to show all your working.

Name	Surface area	Volume
Triangular prism	118 cm^2	60 cm^3
Cylinder	150 cm^2	30 cm^3
Pentagonal prism	112 cm^2	98 cm^3
Cube	148 cm^2	120 cm^3
Cuboid	76.8 cm^2	125 cm^3

3
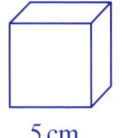
5 cm

6
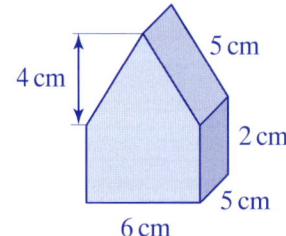

4
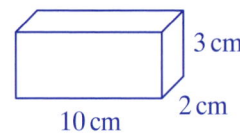

5
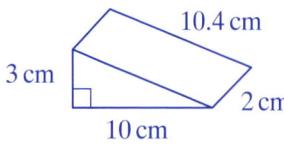

7
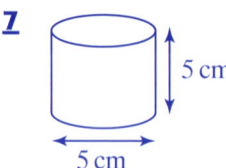

8 Find the surface area of the following cylinders.

Give your answers in terms of π

a   2 cm

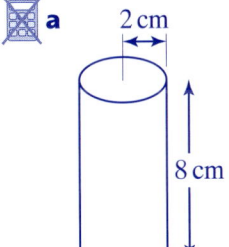

8 cm

b

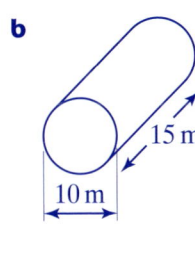

15 m

10 m

2 Equations and inequalities

Solve these equations:

1 $4x = 36$

2 $5y = 65$

3 $3z = 6.9$

4 $4a = 10$

5 $6b = 21$

6 $3c = -12$

7 $x - 1 = 7$

8 $y + 4 = 17$

9 $z + 2.5 = 6.3$

10 $a + 5 = 2$

11 $b - 5 = -6$

12 $c + 4.9 = 2.2$

13 $\dfrac{x}{4} = 6$

14 $\dfrac{y}{11} = 3$

15 $\dfrac{z}{5} = 1.6$

16 $\dfrac{a}{3} = 0.8$

17 $\dfrac{b}{3} = -2$

18 $\dfrac{c}{4} = -1.5$

19 $2x - 15 = 6$

20 $4y + 1 = 49$

21 $3z + 5 = 9$

22 $5a + 21 = -4$

23 $7b + 6 = -15$

24 $8c - 5 = 21$

25 Jon thinks of a number, trebles it and subtracts 4.
The answer is 26.
Write this as an equation.
Solve the equation to find Jon's number.

26 Zoë thinks of a number, multiplies it by 6 and adds 13.
The answer is 61.
Write this as an equation.
Solve the equation to find Zoë's number.

27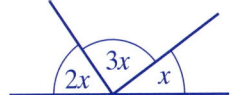

Not drawn accurately

Write down an equation in x.

Solve your equation.

28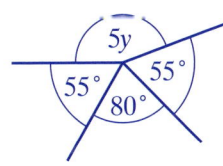

Not drawn accurately

Write down an equation in y.

Solve your equation.

Homework 2

Solve these equations:

1 $4x - 1 = 2x$

5 $p + 2 = 5 - 3p$

9 $3c - 5 = 2 - 4c$

2 $5y - 4 = y + 6$

6 $1 + 2q = 11 - 3q$

10 $2 + d = 5d + 4$

3 $3z - 5 = 10 - 2z$

7 $3 + 7a = 5 - 3a$

11 $12 - 3e = 6 - 5e$

4 $2 - t = 2t + 8$

8 $5b - 10 = 8b - 1$

12 $9f + 7 = 1 + 5f$

13 Jade solves the equation $5x - 3 = 9 - x$ and gets the answer $x = 3$.
Can you find Jade's mistake?

14 Tim solves the equation $4y + 7 = 9 - 2y$ and gets the answer $y = 3$.
Can you find Tim's mistake?

15 $4z - 11 = \blacksquare - z$
The answer to this equation is $z = 4$.
What is the number under the ink blob?

16 $2a + \bullet = 3 - a$
The answer to this equation is $a = -2$.
What is the number under the ink blob?

17 If $b = 3$, find the value of $4b - 7$.
Hence explain why $b = 3$ is not the solution of the equation
$4b - 7 = 4 - 3b$.

18 If $c = -2$, find the value of $9 - 3c$.

Hence explain why $c = -2$ is not the solution of the equation $2c + 11 = 9 - 3c$.

Homework 3

Solve these equations:

1 $3(2x - 5) = 21$

2 $4(y + 3) = 3y$

3 $7 = 2(z - 5)$

4 $2(3p + 1) = 4p$

5 $3(q - 4) = 3 - 2q$

6 $5(2t - 7) = 25$

7 $5a = 2(a - 6)$

8 $2(2b - 9) = 2 + 3b$

9 $2 + 5c = 2(c - 2)$

10 $13d - 5 = 3(3d + 7)$

11 $6(1 - e) = 1 + 4e$

12 $12 - f = 3(4 - f)$

13 $3(7 + 3x) - 25 = 2x + 3$

14 $2(y - 9) + 3(y - 4) = 5$

15 $10 - 2(z + 5) = 8 - 3z$

16 $14 = 5 - 3(2t - 1)$

17 $3(3p - 2) - 4(p + 3) = 7$

18 $3(2q - 7) - (3q - 8) + 22 = 0$

19 Meena thinks of a number, adds 2 and then trebles the result.

Her answer is 36.

Write this as an equation.

Solve the equation to find Meena's number.

20 Jake thinks of a number, subtracts 5 and then multiplies the result by 4.

His answer is 52.

Write this as an equation.

Solve the equation to find Jake's number.

Homework 4

Solve these equations:

1 $\dfrac{x}{2} = 11$

2 $\dfrac{x}{9} = 3$

3 $\dfrac{x}{4} - 1 = 2$

4 $\dfrac{y}{5} + 2 = 7$

5 $\dfrac{z}{8} + 5 = 2$

6 $4 - \dfrac{a}{3} = 1$

7 $7 + \dfrac{b}{2} = 4$

8 $8 = \dfrac{c}{7} - 2$

9 $\dfrac{4x + 1}{3} = 7$

10 $\dfrac{3y - 5}{4} = 4$

11 $\dfrac{4z - 3}{5} = 3 - z$

12 $\frac{1}{4}(2p + 7) = 3$

13 $\frac{1}{2}(3q + 1) = 5$

14 $\frac{1}{3}(4t - 9) = 5$

15 $\frac{1}{2}(3a - 1) = a + 5$

16 $\frac{1}{8}(2b - 7) = 1 - b$

17 $2c - 5 = \frac{1}{3}(6 - c)$

18 $\dfrac{x}{2} + \dfrac{x}{8} = 10$

19 $\dfrac{y}{3} - \dfrac{y}{12} = 7$

20 Dylan says the answer to the equation $\dfrac{x + 4}{5} = 8 - x$ is $x = -9$.

Use substitution to check whether Dylan is correct.

21 Kamala and Jo are solving the equation $\dfrac{2y + 5}{4} = y - 1$.

Kamala gets the answer $y = 2.5$ and Jo gets $y = 4.5$

Check their answers to see if either of them is correct.

22 Explain why you cannot solve the equation $\dfrac{6z - 5}{2} = 4 + 3z$.

Homework 5

1 Write the inequalities shown on these diagrams.

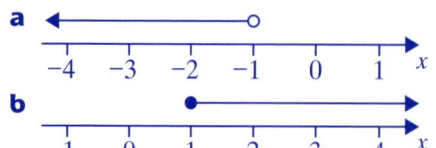

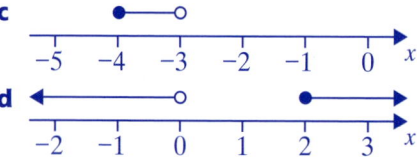

2 List all the integer values of t such that:

    a $-10 < 5t \leqslant 7$ **b** $-9 \leqslant 3t < 3$ **c** $-3 < 2t + 3 \leqslant 5$

3 Find the smallest integer n such that $2n + 7 \leqslant 4n - 2$.

4 Solve these linear inequalities, showing your solutions on a number line.

    a $2x - 3 \geqslant x + 9$ **c** $\dfrac{x}{4} + \dfrac{2x - 5}{6} < 5$

    b $2(3 - 4x) > 1 - 6x$ **d** $\dfrac{x - 2}{3} \leqslant x - \dfrac{2(x + 1)}{5}$

5 Mrs Reynolds says to her class, 'I am thinking of an integer. When I double it and add 5, the result is more than the original number plus 1.'

    a Write the problem as an inequality.

    b What is the smallest number that Mrs Reynolds is thinking of?

6 The integers x and y are such that $-3 < x \leqslant 2$ and $-2 \leqslant y < 1$.

    a What is the smallest value of x^2?

    b What is the largest value of y^2?

    c What is the largest value of xy?

3 Reflections and rotations

Homework 1

1 Copy each shape onto squared paper.

Draw its reflection in the mirror line.

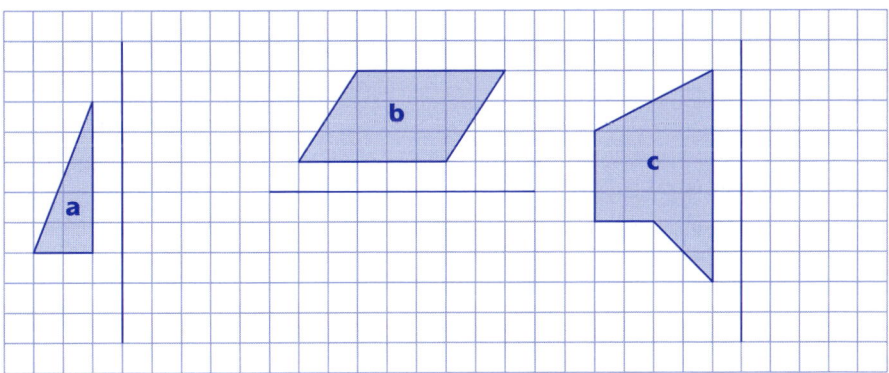

2 Copy these onto squared paper.

Draw the reflection of each shape in the mirror line.

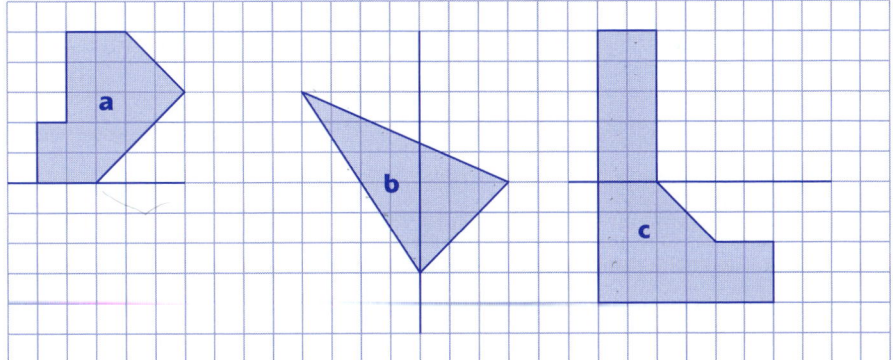

3 Copy each shape onto squared paper.

Draw its reflection in the mirror line.

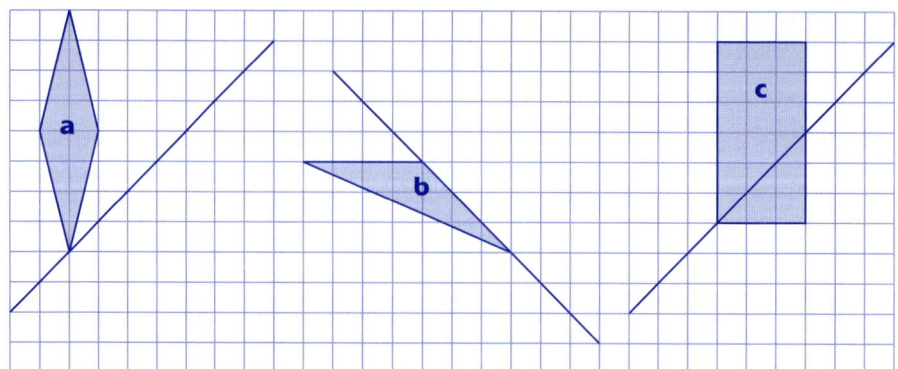

4 a i Copy the diagram onto squared paper.

ii Reflect A in the *x*-axis, and label its image P.

iii Reflect B in the *x*-axis and label its image Q.

b Repeat part **a** on new axes, but now reflect each shape in the *y*-axis.

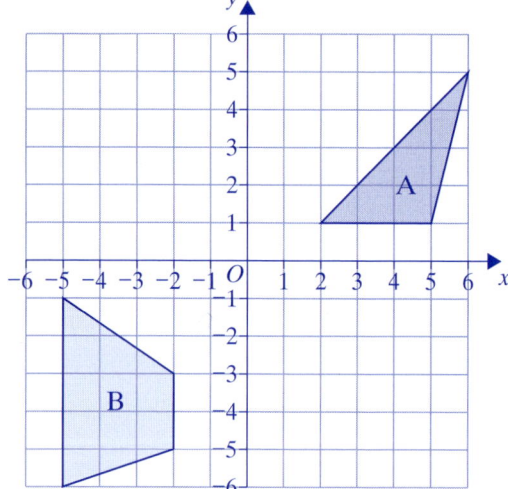

5 For this question use axes of *x* and *y* from −6 to 6.

a i Join the points (0, 0), (−3, 0) and (0, 5) to give a triangle. Label it T.

ii Join (3, −6), (6, −4), (3, −2) and (0, −4) to give a rhombus. Label it R.

iii Draw the reflection of each shape in the *y*-axis.

b Repeat part **a** on new axes, but now reflect each shape in the *x*-axis.

6 Anna says that when point P(1, 5) is reflected in the *y*-axis the image is Q(1, −5).

a Draw a sketch to show these points.

b Is Anna correct? Explain your answer.

7 Use axes of x and y from −8 to 8.

   a Draw a parallelogram, P, with vertices at (0, 3), (3, 0), (3, 4) and (0, 7).

   b Draw the reflection of P in the line $x = 4$ and label it Q.

   c Draw the reflection of P in the line $x = −2$ and label it R.

8 Use axes of x and y from −6 to 6.

   a Draw triangle A by joining (−3, −2), (4, 0) and (2, 2).

   b Draw the reflection of A in the line $y = 2$. Label it B.

   c Draw the reflection of A in the line $y = −2$. Label it C.

   d Mark the sides and angles of A, B and C to show which are equal.

9 When Sam was asked to reflect shape A in the line $y = 2$, he drew this diagram.

   a What did he do wrong?

   b Draw a graph to show the correct image.

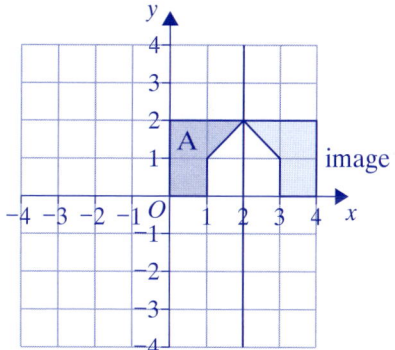

10 **Get Real!**

Many patterns on tiles are formed by reflecting shapes.

This diagram shows half of a tile.

Copy this onto squared paper and reflect the shapes in the line AB to show the rest of the pattern.

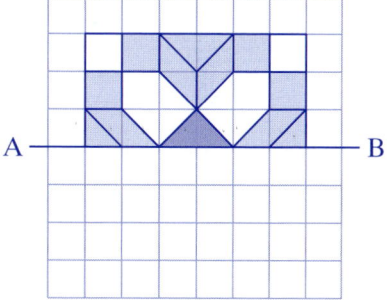

<u>11</u> **a** Draw the mirror line $y = x$ on axes of x and y from −8 to 8.

   b Draw pentagon P with vertices at (1, 4), (4, 4), (4, 6), (2, 7) and (1, 6) and its image, Q, after reflection in the line $y = x$.

   c Draw triangle T with vertices at (2, −2), (−2, −4) and (4, −4) and its image, U, after reflection in the line $y = x$.

12 For this question use axes of x and y from -8 to 8.

    a Draw

        i triangle A by joining $(3, -3)$, $(7, 1)$ and $(8, -4)$

        ii quadrilateral Q by joining $(-3, 7)$, $(2, 7)$, $(3, 5)$ and $(1, 2)$.

    b Draw the image of each shape after reflection in the line $y = -x$.

    c Each shape and its image are congruent.

      Mark the sides and angles to show which are equal to each other.

Homework 2

1 Each of these shapes has one line of symmetry.

Copy each shape onto squared paper and draw its line of symmetry.

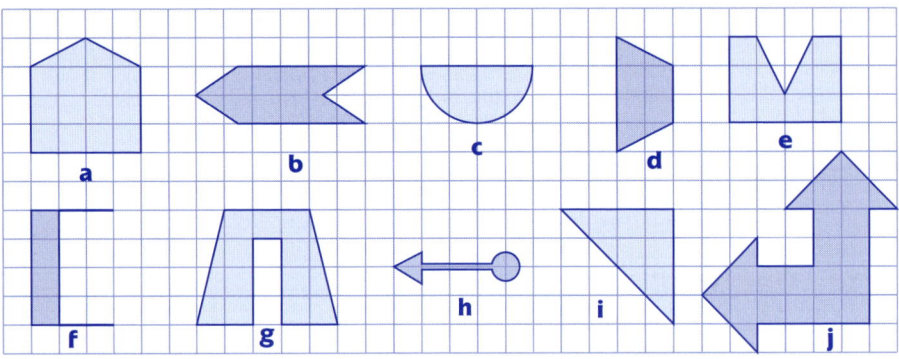

2 Copy these shapes onto squared paper. Draw all their lines of symmetry.

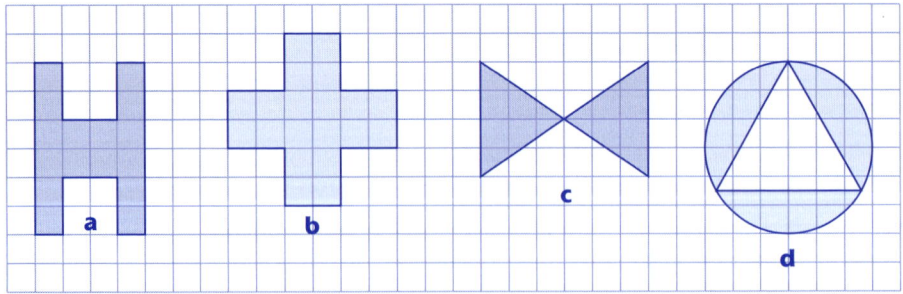

3 Get Real!

How many lines of symmetry does each of these road signs have?

a T-junction

c Dual carriageway ends

b Sharp bend

d Two-way traffic crosses one-way road

4 How many lines of symmetry does each letter in the word **INDEX** have?

5 Copy and complete this table.

Shape	Number of lines of symmetry
Club symbol ♣	
Diamond symbol ♦	
Heart symbol ♥	
Spade symbol ♠	

6 Copy and complete each shape so that the dotted line is a line of symmetry.

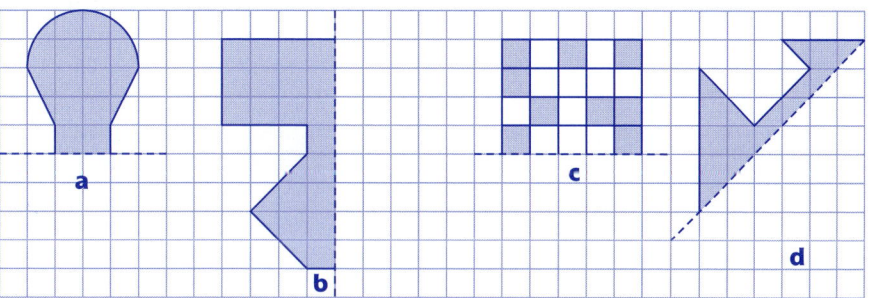

7 **a** Use squared paper for this part.

   i Draw any shape that has two lines of symmetry.

   ii Draw any shape that has four lines of symmetry.

   b Use isometric or isometric dotty paper for this part.

   i Draw any shape that has three lines of symmetry.

   ii Draw any shape that has six lines of symmetry.

8 Which of these 3-D shapes have reflection symmetry?

In each case say how many different ways you can cut the shape into matching halves.

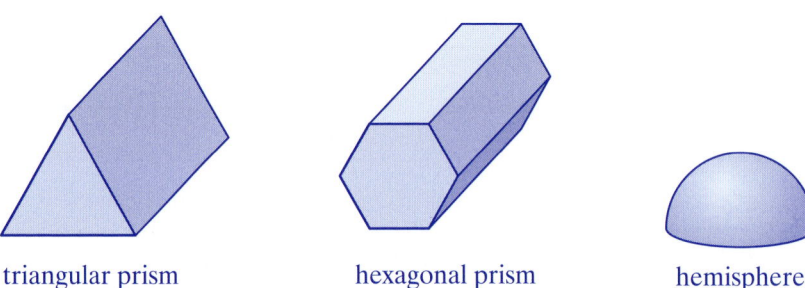

triangular prism   hexagonal prism   hemisphere

9 The following shapes are made from cubes.

How many planes of symmetry does each shape have?

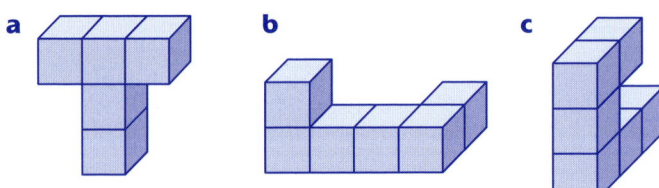

a   **b**   **c**

Homework 3

1   Write down the order of rotation symmetry of each shape.

Use tracing paper to help if you wish.

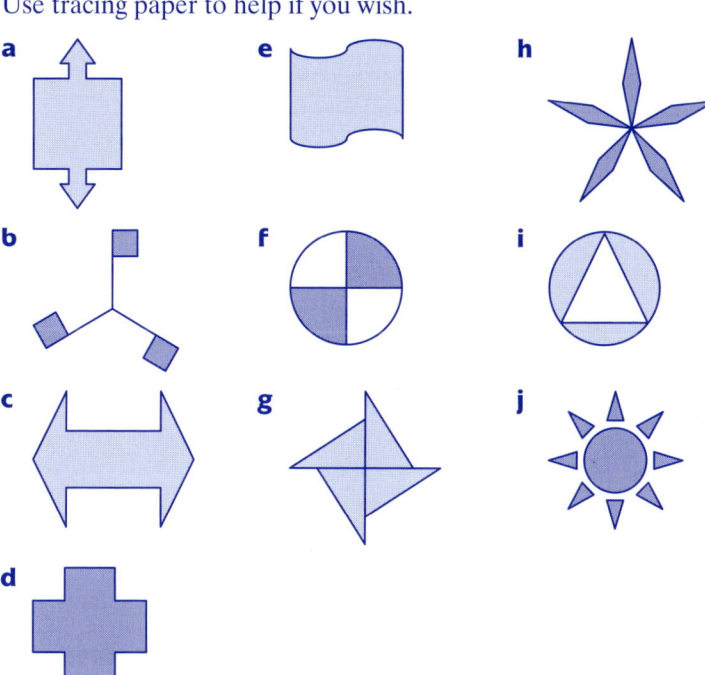

2   Find the order of rotation symmetry of each shape in question **2** of Homework **2**.

3   **Get Real!**
What order of rotation symmetry does each of these diversion signs have?

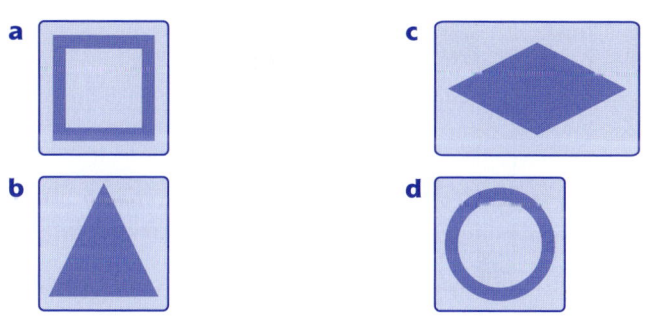

4   What order of rotation symmetry does each letter in the following word have?

INDEX

5 Copy and complete this table.

Shape	Order of rotation of symmetry
Club symbol ♣	
Diamond symbol ♦	
Heart symbol ♥	
Spade symbol ♠	

6 Describe carefully the line and rotation symmetry in each of the following logos.

a 　　b 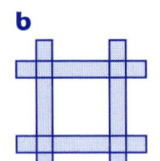　　c 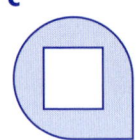　　d 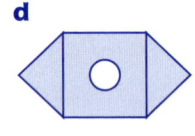　　e

7 **a** On squared paper

  i draw any shape that has rotation symmetry of order 2

  ii draw any shape that has rotation symmetry of order 4.

  b On isometric paper

  i draw any shape that has rotation symmetry of order 3

  ii draw any shape that has rotation symmetry of order 6.

8 **Get Real!**
Crossword puzzles usually have rotation symmetry of order 2.

Copy and complete these crossword patterns so that they have rotation symmetry of order 2.

a 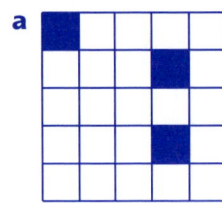　　**b**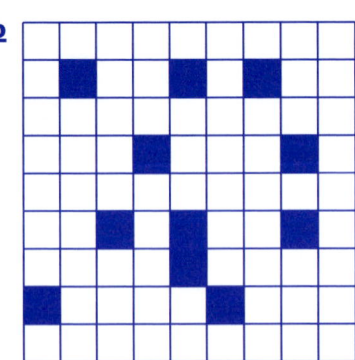

9 **a** Name the special type of triangle that has one line of symmetry and rotational symmetry of order 1.

  b Draw a pentagon that has one line of symmetry and rotational symmetry of order 1.

Homework 4

1 Copy each shape and show its image after the rotation described.

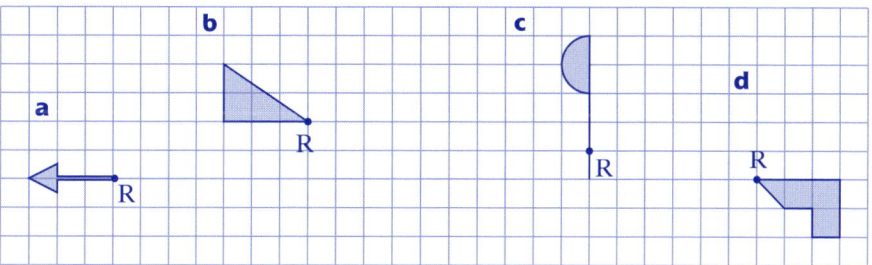

  a Rotate through 90° clockwise about R.

  b Rotate through a half turn about R.

  c Rotate through a quarter turn anticlockwise about R.

  d Rotate through 180° about R.

2 Copy each shape and show its image after the rotation described.

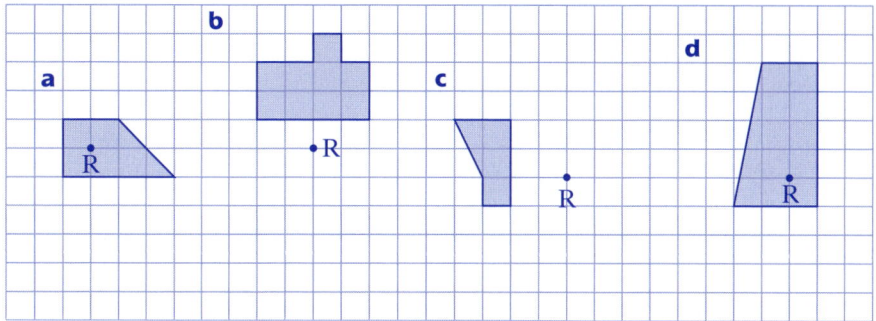

  a Rotate through 90° anticlockwise about R.

  b Rotate through a half turn about R.

  c Rotate through a quarter turn clockwise about R.

  d Rotate through 180° about R.

3 Copy each shape onto isometric paper and show its image after the rotation described.

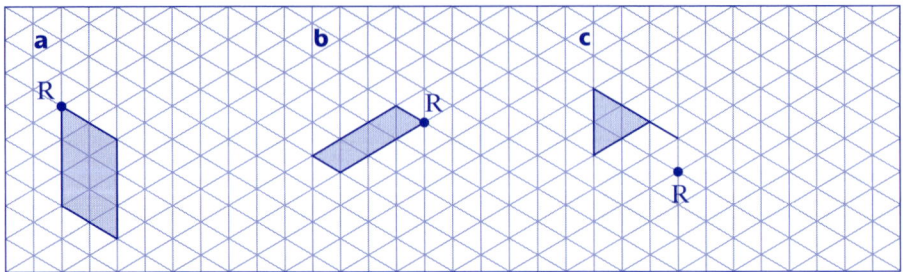

  a Rotate through 60° anticlockwise about R.

  b Rotate through 120° anticlockwise about R.

  c Rotate through $\frac{1}{3}$ turn clockwise about R.

4 **a** On squared paper draw axes of x and y from −6 to 6 and the triangle with vertices at (2, 0), (4, 3) and (2, 3). Label the triangle A.

  b Rotate A through 180° about the origin O and label the image B.

5 **a** On squared paper draw axes of x and y from −6 to 6 and the kite with vertices (0, −3), (2, −2), (4, −3) and (2, −6). Label the kite K.

  b Rotate K through a half turn about the origin O and label the image L.

6 **a** Copy the diagram onto squared paper.

  b Draw the image of the Z-shape after a rotation of 90° anticlockwise about the origin O. Label it A.

  c Draw the image of the Z-shape after a rotation of 180° about the origin O. Label it B.

  d Draw the image of the Z-shape after a rotation of 90° clockwise about the origin O. Label it C.

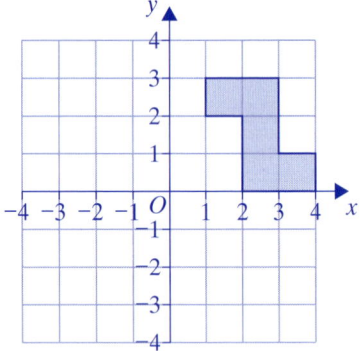

7 **a** On axes of x and y from −6 to 6, draw triangle T with vertices (−4, 2), (0, 2) and (−1, 5).

  b Draw the image of T after a rotation of 90° clockwise about O and label the image U.

8 **a** On axes of x and y from −6 to 6, draw parallelogram P by joining (0, −2), (4, −2), (6, −4) and (2, −4).

  b Draw the image of P after a quarter turn rotation clockwise about O and label the image Q.

9 Meera says that when point (−2, 4) is rotated 90° anticlockwise about the origin the image point is (2, −4). Draw a sketch and say whether or not Meera is right.

10 **Get Real!**

The diagram shows the end view of a window.

From this position it can be opened by rotating RS up to 50° clockwise about R.

Copy RS and show its position when it is fully open.

11 On squared paper draw axes of x and y from −8 to 8.

a Draw and label trapezium T with vertices (3, 4), (5, 4), (6, 6) and (0, 6).

b Draw the image of T after a half turn about the point A(0, 4) and label it U.

c Draw the image of T after a half turn about the point B(0, −1) and label it V.

d Draw the image of T after a half turn about the point C(4, −1) and label it W.

e What do you notice about the three images?

12 On squared paper draw axes of x and y from −6 to 6.

a Join the points (2, −2), (3, −6) and (6, −6) to form a triangle and label it A.

b Draw the image of A after a rotation of 90° clockwise about (2, −2). Label it B.

c Draw the image of A after a rotation of 90° clockwise about the point (5, 1). Label it C.

d Mark the sides and angles in A, B and C to show which are equal.

e What can you say about the images?

Homework 5

1 Copy these objects and their images onto squared paper.

In each case draw the mirror line.

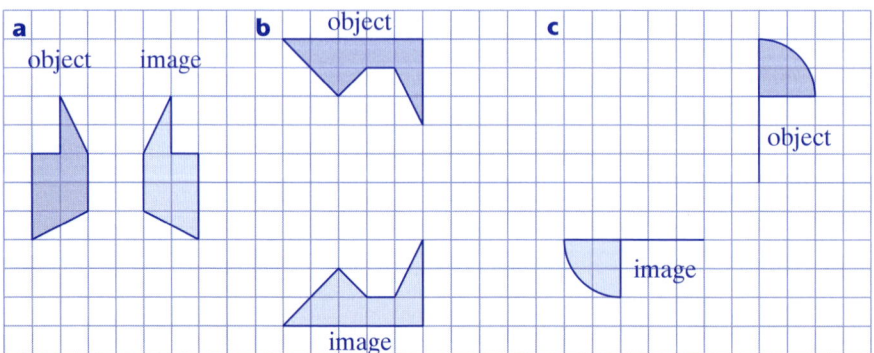

2 Find the equation of the mirror line that maps:

a A onto B

b A onto C

c C onto D

d B onto E

e D onto H

f E onto F

g G onto K

h F onto I

i G onto E

j J onto I

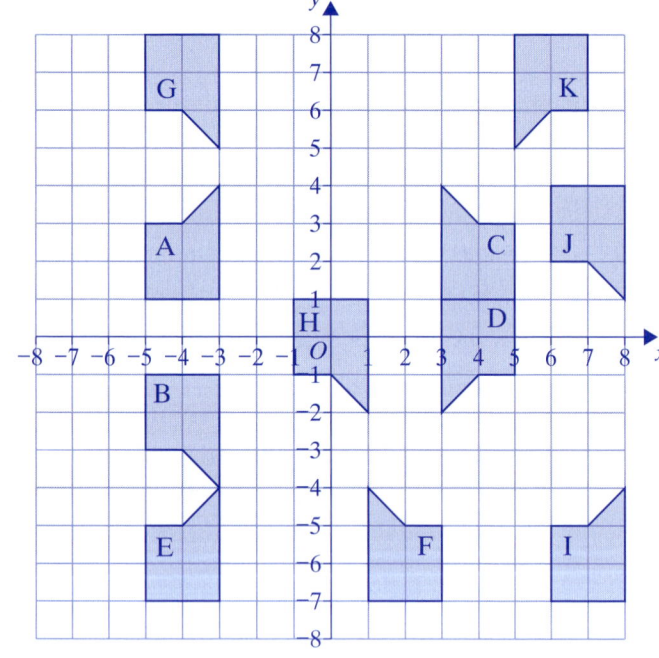

3 **a** On axes of x and y from −8 to 8 draw and label triangle A with vertices (−3, 6), (−3, 4) and (0, 4).

    b **i** Draw and label triangle B with vertices (−6, 3), (−4, 3) and (−4, 0).

      ii Describe fully the transformation that maps A onto B.

    c **i** Draw and label triangle C with vertices (6, −3), (4, −3) and (4, 0).

      ii Describe fully the transformation that maps A onto C.

    d Describe fully the transformation that maps B onto C.

4 For each part the diagram shows an object and its image after a rotation about R.

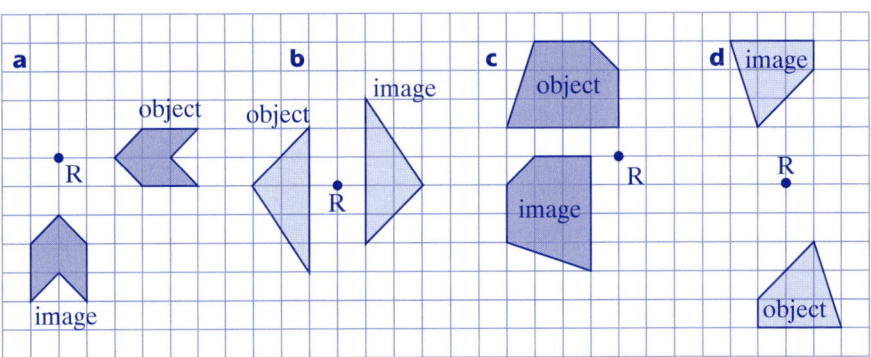

In each case give the angle of rotation and state whether it is clockwise or anticlockwise.

5 Describe fully the rotation that maps:

    a K onto L

    b K onto M

    c K onto N

    d L onto N

    e L onto K

    f M onto K

    g M onto N

    h N onto M

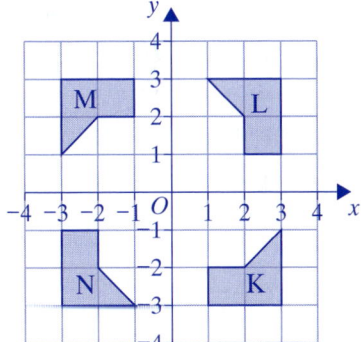

6 Describe fully the rotation that maps:

   a A onto B

   b A onto C

   c A onto D

   d D onto E

   e B onto D

   f A onto F

   g C onto G

   h F onto H

   i B onto E

   j C onto D

   k H onto A

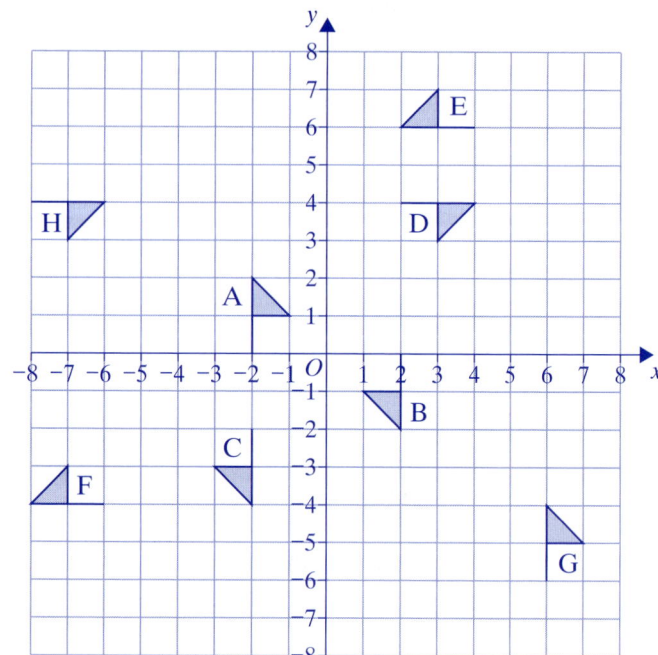

7 Describe fully the transformation that maps:

   a A onto B

   b B onto C

   c A onto D

   d C onto D

   e A onto C

   f D onto B

   g C onto B

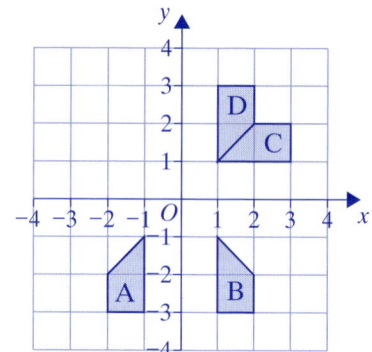

8 Get Real!

   a Describe fully the transformation that maps the minute hand on a clock from its position at twelve o'clock to its position at quarter past twelve.

   b Describe fully the transformation that maps the hour hand on a clock from its position at six o'clock to its position at eight o'clock.

9 The diagram shows a quadrilateral ABCD and its image PQRS after a transformation.

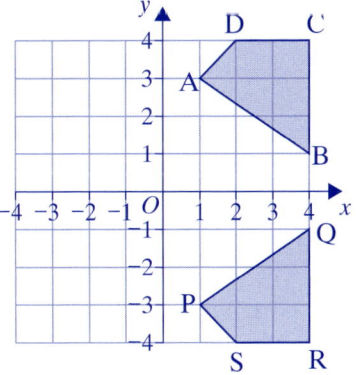

a Give a full description of the transformation.

b Find and name the length that is equal to:

   i AB   **ii** CD   **iii** BD   **iv** AC

c Find and name the angle that is equal to:

   i ∠CDA   **ii** ∠CAD   **iii** ∠AOB

d Find and name a triangle that is congruent to:

   i DAB   **ii** CAD

10 **a** On axes of x and y from −6 to 6 draw rhombus R with vertices (4, 2), (5, 4), (4, 6) and (3, 4).

  b **i** Draw and label rhombus S with vertices (−4, 2), (−3, 4), (−4, 6) and (−5, 4).

     ii Find as many transformations as you can that map R onto S.

     Describe each transformation fully.

  c **i** Draw and label rhombus T with vertices (−4, −2), (−5, −4), (−4, −6) and (−3, −4).

     ii Find as many transformations as you can that map R onto T.

     Describe each transformation fully.

  d Find as many transformations as you can that map T onto S.

  Describe each transformation fully.

Homework 6

1 Use squared paper for this question.

a Draw four parallel mirror lines.

b Draw a triangle in the centre as shown.

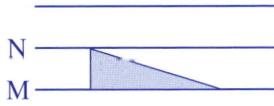

c Reflect the triangle in each mirror line it touches (that is, in mirror lines M and N).

d Reflect the images in each mirror line they touch.

2 **a** Draw two perpendicular mirror lines M and N that intersect at R and an F-shape, A, like this on squared paper.

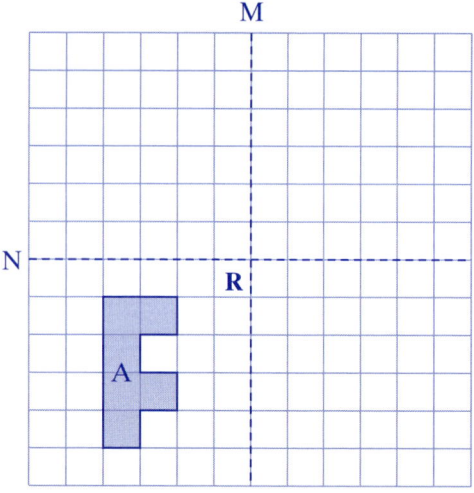

b Reflect A in mirror line M.

Label the image B.

c Now reflect B in mirror line N.

Label the image C.

d Describe fully the transformation that would map A onto C.

3 **a** Draw another copy of the diagram in question **2**.

b Rotate A through 90° anticlockwise about R. Label the image B.

c Now rotate B through 180° about R. Label the image C.

d Describe fully the transformation that would map A onto C.

4 **a** Using axes of x and y from −6 to 6, draw and label triangle T with vertices (−4, 5), (−2, 5) and (−4, 2).

b Draw the reflection of triangle T in the y-axis and label it U.

c Now draw the reflection of U in the x-axis and label it V.

d Describe fully the single transformation that would map triangle T onto V.

5 **a** Using axes of x and y from −6 to 6 draw trapezium A with vertices at (3, 4), (4, 4), (6, 2) and (2, 2).

b Rotate A through 180° about O and label the image B.

c Now reflect B in the x-axis and label the image C.

d Describe fully the single transformation that would map A onto C.

6 **a** Using axes of x and y from −6 to 6 draw kite K by joining (3, 5) (2, 4) (3, 1) and (4, 4).

b Rotate K through 90° clockwise about O and label the image L.

c Now draw the reflection of L in the y-axis and label it M.

d Describe fully the single transformation that would map K onto M with point (2, 4) moving to point (−4, −2).

7 Use axes of x and y from -8 to 8.

    a Draw pentagon P with vertices at $(1, 4)$, $(4, 4)$, $(4, 7)$, $(3, 8)$ and $(1, 7)$.

    b Draw the reflection of P in the line $y = x$ and label the image R.

    c Now draw the reflection of R in the x-axis and label it S.

    d Describe fully the single transformation that would map P onto S.

8 Use axes of x and y from -6 to 6.

    a Join the points $(0, 3)$, $(5, 4)$, $(4, 1)$ and $(2, 1)$ to give a quadrilateral, A.

    b i Rotate A through 180° about $(3, 0)$ and label the image B.

      ii Mark the sides and angles in A and B to show which are equal.

    c i Now rotate B through 90° anticlockwise about $(-2, -2)$ and label the image C.

      ii Mark the sides and angles in C to show which are equal to those in A and B.

    d Describe fully the single transformation that would map A onto C.

9 Sufia says, 'If you reflect a shape in $y = x$ and then rotate it 90° anticlockwise about O, you always get the same image that you would get if you rotated it rotated it 90° anticlockwise about O then reflected it in $y = x$.'

Use the triangle with vertices at $(3, 1)$, $(6, 1)$ and $(3, 3)$ to check whether or not this statement is true.

10 On squared paper draw axes of x and y from -8 to 8.

    a Draw a parallelogram by joining $A(1, 6)$, $B(4, 6)$, $C(2, 4)$ and $D(-1, 4)$.

    b Draw the reflection of ABCD in the line $y = x$ and label the image $A_1 B_1 C_1 D_1$ where A_1 is the image of A, B_1 is the image of B and so on.

    c Now draw the reflection of $A_1 B_1 C_1 D_1$ in the line $y = -x$ and label it $A_2 B_2 C_2 D_2$ where A_2 is the image of A_1, B_2 is the image of B_1 and so on.

    d Describe fully the single transformation that would map ABCD onto $A_2 B_2 C_2 D_2$ with A mapped to A_2, B to B_2 etc.

    e Describe in words another way of moving ABCD onto $A_2 B_2 C_2 D_2$ but in which A would not move to A_2.

11 What single transformation is equivalent to reflection in $y = x$ followed by reflection in $y = 2$?

Draw a diagram to illustrate your answer.

4 Ratio and proportion

Homework 1

 1 Simplify these ratios.

a 3 : 6		**e** 3 : 18		**i** 25 : 35		**m** 0.6 : 0.4	
b 3 : 9		**f** 3 : 21		**j** 75 : 100		**n** $3\frac{1}{2} : 10\frac{1}{2}$	
c 3 : 12		**g** 14 : 28		**k** $\frac{2}{3} : \frac{5}{6}$		**o** 25% : 75%	
d 3 : 15		**h** 9 : 36		**l** 2.5 : 3.5		**p** 75 : 200	

 2 **a** Write down three different pairs of numbers that are in the ratio 3 : 1.

    b Write down three different pairs of numbers that are in the ratio 5 : 1.

    c Explain how to find pairs of numbers that are in the ratio 5 : 1.

    d Harry writes the three pairs of numbers 16 and 20, 14 and 18, and 12 and 16. He says these pairs of numbers are all in the same ratio. Is Harry correct? Explain your answer.

 3 Get Real!

Ben's GCSE results are D, A, B, C, C, E, D, B and C.

What is the ratio of Ben's number of Grade C and above results to his number of below grade C results?

 4 Get Real!

In a large college, staffing costs are £15 million and other costs are £5 million. Find the ratio of staffing costs : other costs in its simplest form.

 5 Get Real!

A recipe lists the following quantities of ingredients to make six scones.

Self-raising flour	240 grams
Salt	0.5 teaspoon
Sultanas	75 grams
Butter	40 grams
Caster sugar	25 grams
Egg	1 large
Milk	20 ml

    a List the ingredients needed for 12 scones.

    b How much flour would be needed for 15 scones?

6 Get Real!
In a nursery, the number of children is 24 and the number of nurses is 4.

a What is the nurse : child ratio?

b Using this ratio, find how many nurses are needed for 30 children.

 7 Get Real!
Red and yellow paint are mixed in the ratio 2 : 5 to make the colour 'Hot Orange'.

a Which of these mixes will make 'Hot Orange' paint?

Amount of red paint (litres)	Amount of yellow paint (litres)
2	5
3	6
3	4
5	2
4	10
6	15

b Write down another possible pair of amounts of red and yellow paint that would make 'Hot Orange'.

c Would it make any difference if the amounts were measured in pints instead of litres? Explain your answer.

8 Get Real!
The aspect ratio of a film is the ratio of the width of the picture to its height. A common Cinemascope aspect ratio is 2.35 : 1.

Using this aspect ratio, find:

a the width of a picture if its height is 2 metres

b the height of a picture whose width is 4 metres.

9 The ratio $x:y$ simplifies to 2 : 9.

a If x is 6, what is y?

b If y is 36, what is x?

c If y is 13.5, what is x?

d If x and y add up to 55, what are x and y?

10 Make up another question like question **9** and give the answers.

11 **Get Real!**

a Find, in their simplest forms, the teacher : student ratios for these schools.

School	Number of teachers	Number of students
School 1	63	945
School 2	16	280
School 3	27	567
School 4	125	2275
School 5	55	880

b i If a school with 40 teachers had the same teacher : student ratio as School 1, how many students would it have?

ii If a school with 1500 students had the same teacher : student ratio as School 1, how many teachers would it have?

Homework 2

1 Divide these numbers and quantities in the ratio 1 : 4.

a 200   **c** £6.50   **e** £2.50

b 350   **d** 8 litres   **f** 3.5 litres

2 Divide the numbers and quantities in question **1** in the ratio 2 : 3.

3 Divide the numbers and quantities in question **1** in the ratio 7 : 3.

4 Divide the numbers and quantities in question **1** in the ratio 1 : 4 : 5.

5 **Get Real!**

Salad dressing is made from oil and vinegar in the ratio 3 : 1.

a How much oil is needed to make 100 mℓ of salad dressing?

b How much vinegar is needed to make 0.4 litres of salad dressing?

c How much salad dressing can you make if you have plenty of oil but only 20 mℓ of vinegar?

6 a Find the number of boys and the number of girls in these schools.

School	Total number of students	Boy : Girl ratio
School A	2000	1 : 1
School B	1800	2 : 7
School C	1860	2 : 3
School D	525	8 : 7
School E	1055	100 : 111

b Find the boy : girl ratios in part **a** in the form $1 : n$ (in other words, find how many girls there are for every boy).

c Which school has the largest proportion of girls? Give a reason for your answer.

7 a Work out the amount of protein in 100 g of each of these foods.

Food	Carbohydrate:fat:protein ratio
Fudge biscuits	9:2:1
Strawberries	11:1:1
Scambled eggs	1:2:2
Chilli con carne	7:6:7
Italian sausage	1:5:4

b How many grams of chilli con carne do you need to eat to have 100 g of protein?

c How many grams of fudge biscuits do you need to eat to have 100 g of protein?

8 A school has £10 000 to spend on ICT equipment. The amount is to be split between the lower school and the upper school in the ratio of the number of students. There are 545 students in the lower school and 495 in the upper school.

How much money does each part of the school have to spend?

9 Pen metal for pen nibs is an alloy of copper, gold and silver in the ratio 2 : 1 : 1.

a How much of each metal is needed to make 1 kilogram of pen metal?

b If there are only 145 grams of gold left but plenty of copper and silver, how much pen metal can be made?

c How much gold would there be in a pen nib weighing 2 grams?

Homework 3

1 Get Real!

Jane pays £2.40 for 8 pencils.

How much would 20 of the same pencils cost?

2 Get Real!

Ali worked for 6 hours one day and earned £105.60

  a How much will he be paid for a day when he works for 8 hours at the same rate of pay?

  b Complete a copy of this table. Plot the values in the table as points on a graph, using the numbers of hours worked as the x-coordinates and the money earned as the corresponding y-coordinates.

Number of hours worked	0	2	4	6	8	10
Money earned (£)				105.60		

  c The points should lie in a straight line through (0, 0).

    i Explain why.

    ii What does the gradient of the line represent?

    iii Show how to use the graph to find out how long Ali has to work to earn £100.

3 Get Real!

To make sandwiches for 3 people on 5 days, 750 g of meat is needed.

  a How much meat is needed for sandwiches for 8 people on 1 day?

  b How many days' sandwiches for 4 people will 600 g of meat make?

  c Sandwiches for 5 people for 4 days needs the same amount of meat as sandwiches for 10 days for how many people?

4 Get Real!

Softouch hand cream is packed in small and large sizes, as shown in the table.

	Amount of hand cream	Price
Small size	55 grams	£1.05
Large size	225 grams	£4.35

Which size is better value for money?

Show how you worked out your answer.

5 Get Real!

A journey of 240 kilometres took 5 hours.

a How far would you go in 4 hours at the same average speed?

b How far would you go in three quarters of an hour at this average speed?

c How long would it take to travel 400 kilometres at this average speed?

 **6 Get Real!**

A mail order company offers to supply and deliver boxes of

22 chocolates for £17.95

44 chocolates for £27.95

66 chocolates for £35.95

a Which of these is best value for money? Show how you worked it out.

b Write a sentence suggesting why there is such a difference in price per chocolate.

7 Get Real!

Notice that the two parts of this question are really the same! Use part **a** to help you work out part **b**.

a 75% of a number is 6. Use the unitary method to find 100% of the number.

b A T-shirt is £6 with the special offer (25% off). What was the original price of the T-shirt?

8 a Two numbers are in the ratio 1 : 0.6

The first number is 15; what is the second?

b Two numbers are in the ratio 1 : 0.6

The second number is 12; what is the first?

c Three numbers are in the ratio 1.2 : 1 : 0.8

The third number is 36; what are the other two numbers?

9 Get Real!

The weights of objects on other planets are proportional to their weights on Earth. A person weighing 150 pounds on Earth would weigh 60 pounds on Mercury and 160 pounds on Saturn.

a What would a person weighing 100 pounds on Earth weigh on Mercury?

b What would a mineral sample weighing 20 kilograms on Saturn weigh on Earth?

c Sketch a graph to show the weights of objects on Mercury compared with their weights on Earth.

d Express the ratio 'weight of object on Earth : weight of object on Mercury : weight of object on Saturn' as simply as you can.

Homework 1

1 Decide which of the following statements are true and which are false.

   a 5.2^2 is between 25 and 36.

   b 9.11^2 is between 18 and 20.

   c 8.8^2 is between 64 and 81.

   d 8.9^2 is between 64 and 81.

   e 0.5^2 is between 0 and 1.

   f 1.99^3 is between 1 and 8.

   g 10.3^3 is between 100 and 121.

2 Use the trial and improvement method to solve these equations.

   Give your answers to one decimal place.

   Remember to show all your working.

   a $x^2 = 72$ **b** $x^2 = 44$ **c** $x^3 = 105$ **d** $x^3 = 54$

3 Use the trial and improvement method to solve the following equations.

   Give your answers to one decimal place.

   Remember to show all your working.

   a $x^2 + 10 = 45$ **c** $x^2 - 11 = 100$ **e** $x^2 - 312 = 800$

   b $x^2 + 1 = 100$ **d** $x^2 - 100 = 25$

4 The area of this square is 110 cm^2.

Area 110 cm^2

   Use the trial and improvement method to find the length of the square.

   Give your answer to two decimal places.

5 The volume of a dice is 4 cm^3.

   Use the trial and improvement method to find the length of a side of the dice.

   Give your answer to two decimal places.

6 Use the trial and improvement method to find the length of this rectangle to two decimal places.

Remember to show all your working.

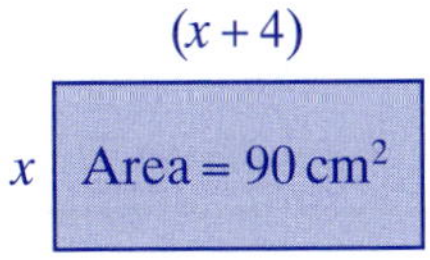

$(x+4)$

x | Area = 90 cm²

7 The area of Amir's garden is 100 square metres.

The length of the garden is two metres longer than the width.

Use the trial and improvement method to work out the length and width of the garden.

Give your answers to one decimal place.

8 A solution of the equation $x^3 - 8x = 110$ lies between $x = 5$ and $x = 6$.

Use the trial and improvement method to find this solution.

Give your answer to one decimal place.

9 Use the trial and improvement method to find solutions to these equations.

Give your answers to two decimal places.

a $x^2 + 3x = 46$ if the solution lies between 5 and 6

b $x^3 + 2x = 38$ if the solution lies between 3 and 4

c $x^3 + 21 = 0$ if the solution lies between −2 and −3

Homework 1

1 a Copy and enlarge these shapes, making every line twice as long.

    i 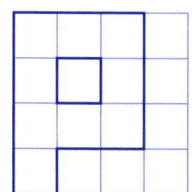   **ii**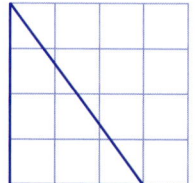

    b Work out the areas of the original and the enlarged shapes.

    c What do you notice?

2 Copy and enlarge this shape by scale factor:

    a 3   **b** $\frac{1}{2}$ **c** 1.5

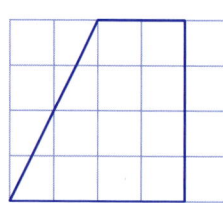

<u>3</u> This rectangle is enlarged with a scale factor 2.5
What are the measurements of the enlarged rectangle?

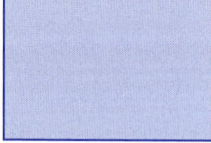

6 cm

9 cm

<u>4</u> A rectangle has been enlarged. The original rectangle has length 20 cm.
The enlarged rectangle has length 25 cm.

    a What is the scale factor of enlargement?

    b If the width of the enlarged rectangle is 50 cm, what is the width of the original rectangle?

    c Find the areas of the two rectangles.

Homework 2

1   For each part, copy the diagram onto squared paper and draw an enlargement with scale factor 2.

Use C as the centre of enlargement.

a
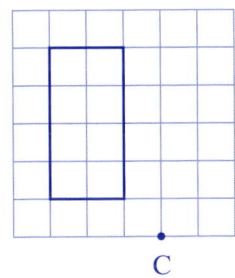

b
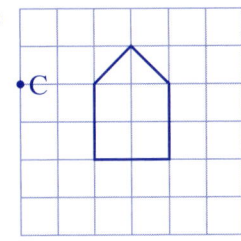

2   For each part, copy the diagram onto squared paper and draw an enlargement with scale factor $\frac{1}{2}$

Use C as the centre of enlargement.

a
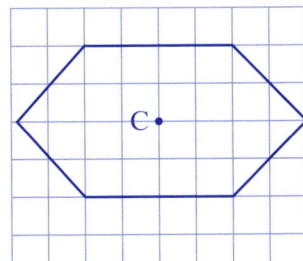

b
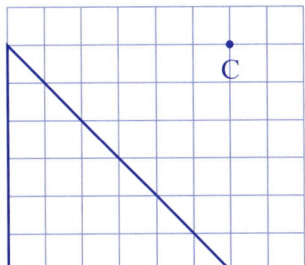

3   In each diagram below, shape A has been enlarged to make shape B.

In each case find the scale factor and the centre of the enlargement.

a
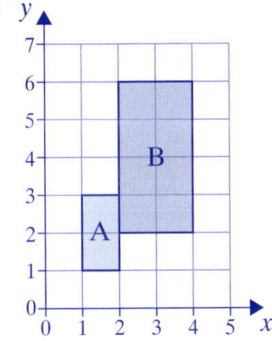

b
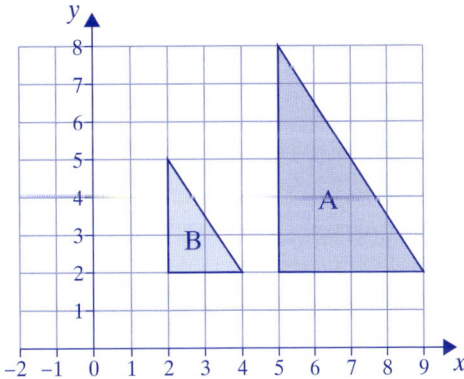

4 Draw axes from −8 to 8.

Draw triangle D using the coordinates (1, −2), (7, −2) and (−3, 8).

a Draw the image of D after an enlargement scale factor of $\frac{1}{2}$, centre (−5, −8). Label the image E.

b What are the coordinates of the vertices of E?

Homework 3

1 Triangle B is an enlargement of triangle A.

a What is the scale factor of the enlargement?

b What is the ratio of the lengths of their bases?

c What is the ratio of their longest sides?

d What is the ratio of the perimeters?

2 Triangle FGH is an enlargement of triangle CDE.

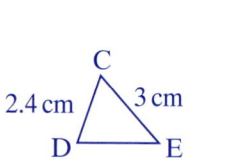
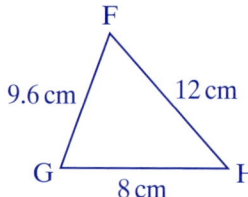

a What is the scale factor of the enlargement?

b What is the length of DE?

c What is the ratio of the perimeters of these triangles (in its simplest form)?

3 Rebecca says that triangle K is an enlargement of triangle J. Is she correct? Give a reason for your answer.

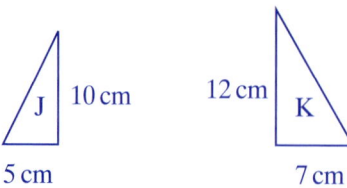

4 Rectangle M is an enlargement of rectangle N.

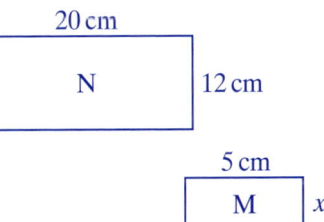

a What is the ratio of the corresponding sides?

b What is the scale factor of the enlargement?

c What is the length of the side marked x?

d What is the ratio of the perimeters of these rectangles?

5 Triangle Q is an enlargement of triangle P.

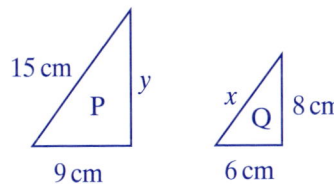

a What is the ratio of the corresponding sides?

b What is the length of the sides x and y?

c What is the ratio of the perimeter of these triangles?

d What is the ratio of the area of these triangles?

Homework 4

1 Write the coordinates of the images of these points after the translations described.

a (2, 4) translation 3 units to the right followed by 1 unit up

b (3, 6) translation 4 units to the right followed by 6 units down

c (−1, 0) translation 3 units to the right followed by 2 units up

d (−1, 3) translation 3 units to the left followed by 4 units down

e (−2, −9) translation 3 units to the left followed by 4 units up

2 Find the coordinates of the image of each of these points after a translation by the given vector.

a (3, 4) by $\begin{pmatrix} 11 \\ 7 \end{pmatrix}$ **c** (0, 4) by $\begin{pmatrix} 7 \\ -8 \end{pmatrix}$

b (2, 0) by $\begin{pmatrix} 3 \\ 4 \end{pmatrix}$ **d** (−2, −3) by $\begin{pmatrix} -5 \\ -4 \end{pmatrix}$

3 Draw axes for both x and y between 0 and 10.

Draw triangle A using the coordinates (1, 3), (1, 7) and (3, 3).

a Draw the image of this triangle after it has been translated 2 units to the right and 3 units up.

Label this triangle B.

b Draw the image of triangle A after it has been translated 3 units to the right and 2 units down.

Label this triangle C.

c Describe the translation that would move triangle B onto triangle C.

4 Draw axes for both x and y between 0 and 6.
Draw triangle X using the coordinates (2, 3), (4, 5) and (5, 1).

a Draw the image of this triangle after it has been translated by $\begin{pmatrix} -2 \\ -1 \end{pmatrix}$.

Label this triangle Y.

b Draw the image of Y after it has been translated by $\begin{pmatrix} 0 \\ 2 \end{pmatrix}$. Label this triangle Z.

c What are the coordinates of triangle Z?

5 Triangle ABC is translated onto triangle A′B′C′. A and B are the points (2, 2) and (7, 2) respectively. A′ and C′ are the points (−3, −5) and (−3, −1) respectively.
Find the coordinates of C and B′.

Homework 5

1 Draw axes for x between −4 and 8 and y between −2 and 5.

Plot the quadrilateral W using the coordinates (3, 2), (4, 2), (4, 4) and (3, 4).

a Translate the quadrilateral 4 units to the right and 0 units up. Label this quadrilateral X.

b Reflect X in the line $x = 2$. Label this quadrilateral Y.

c Describe the single transformation that maps W directly onto Y.

2 Draw axes for x between −4 and 8 and y between −2 and 5. Plot quadrilateral Z with the coordinates (3, 2), (4, 2), (4, 4) and (3, 4).

a Reflect Z in the line $x = 2$. Label this quadrilateral A.

b Translate quadrilateral A using the vector $\begin{pmatrix} 4 \\ 0 \end{pmatrix}$. Label this quadrilateral B.

c Describe the single transformation that maps Z directly onto B.

3 Draw axes from −6 to 6.

Draw triangle T with vertices at (1, 1), (4, 1) and (4, 5).

a Reflect T in the *x*-axis and label the image U.

b Translate U using the vectors $\begin{pmatrix} -5 \\ -1 \end{pmatrix}$ and label the image V.

c Reflect V in the *x*-axis and label the image W.

d Describe the single transformation that maps T directly onto W.

Homework 6

1 Find the distance that 1 cm represents on a map with a scale of:

a 1 : 20   **b** 1 : 500   **c** 1 : 800   **d** 1 : 50 000

2 A map has a scale of 1 : 25 000. Find the actual distance (in kilometres) that each of these lengths on a map represents.

a 2 cm   **b** 3 cm   **c** 5 cm   **d** 6.5 cm

3 Mr Bradfield wants to draw an accurate scale drawing of his garden. He draws a rough sketch.

The garden is 25 m long and is 18 m wide.

The path is 1 m wide.

The lawn is 12 m wide.

The flower beds are 8 m wide.

Draw an accurate scale drawing of the garden using a scale of 1 : 200.

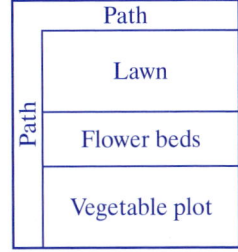

4 On a map with a scale of 1 : 10 000, the distance between Richard's house and his school is 2.5 cm.

What is the actual distance between Richard's house and his school?

5 On the same map the distance between Priya's house and the supermarket is 5 cm.

What is the actual distance between Priya's house and the supermarket?

6 The distance between David's house and the sports centre is 400 m. How far would this be on a map with a scale of 1 : 5000?

7 In the atlas Anwar finds the distance between London and Rome is 6.4 cm. The actual distance is 1408 km. What is the scale of the map in the atlas?

Homework 1

 1 **a** Convert to metres:

    i 4960 mm   **ii** 140 cm   **iii** 0.175 km

   b Convert to kilograms:

    i 3.25 t   **ii** 840 g   **iii** 500 mg

   c Convert to litres:

    i 35 cℓ **ii** 4700 mℓ **iii** 125 mℓ

 2 **a** In the imperial system 1 gallon = 8 pints.

    Write in pints:

    i 2 gallons   **ii** $4\frac{1}{2}$ gallons

   b In the imperial system 1 pound = 16 ounces and 1 stone = 14 pounds.

    Write in pounds:

    i 3 stones   **ii** 48 ounces   **iii** $2\frac{1}{2}$ stones

   c In the imperial system 1 foot = 12 inches, 1 yard = 3 feet and 1 mile = 1760 yards.

    Write in feet:

    i 120 inches   **ii** 8 yards   **iii** a third of a mile

 3 **Get Real!**

The table gives the dimensions of some furniture in centimetres.

Copy the table, but give the dimensions in metres.

Item	Width	Height	Depth
Computer desk	120 cm	72 cm	49 cm
Bookcase	87 cm	123 cm	28.5 cm
Filing cabinet	37.5 cm	130 cm	62 cm
Cupboard	80 cm	81.7 cm	46 cm

4 Copy and complete the following, giving your answers to 2 significant figures. The conversion factors are given in brackets.

a 8 inches = ... cm (1 inch = 2.54 cm)

b 15 feet = ... m (1 m = 3.28 feet)

c 29 miles = ... km (1 mile = 1.61 km)

d 56 pounds = ... kg (1 kg = 2.2 pounds)

e 6 pints = ... ℓ (1 ℓ = 1.76 pints)

f 28 gallons = ... ℓ (1 gallon = 4.55 ℓ)

5 Say which of the following are wrong and why.

a 2.3 m = 230 cm   **d** 18 mg = 0.18 g

b 82 cm = 8.2 mm   **e** 800 mℓ = 80 cℓ

c 750 g = 0.75 kg   **f** 6750 cℓ = $6\frac{3}{4}$ ℓ

6 Get Real!

The Joneses have weighed their cases before going on holiday.

Convert each mass to kilograms to 1 d.p.

10 lb   14 lb   37 lb   54 lb

| lb means 'pounds'; use 1kg = 2.2 lb |

7 Get Real!

The dimensions of a kitchen unit are:

height 900 mm   width 500 mm   depth 600 mm

a Convert each of these dimensions to:

   i cm   **ii** m

   iii inches (use 1 inch = 2.54 cm).

b How many of these units would fit along a wall that is:

   i 3 m long

   ii 7 feet 6 inches long (use 1 foot = 12 inches)?

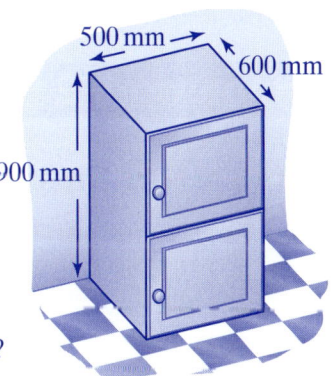

500 mm   600 mm   900 mm

8 Get Real!

Here is an old recipe that makes two pints of tomato soup.

A caterer wants to make enough for 80 servings, each of 150 mℓ, for a school lunch.

Work out how much he needs of each ingredient.

Give answers to the nearest 0.5 litre, 0.1 kg, tablespoon or teaspoon. (Use 1 litre = 1.76 pints, 1 ounce = 28.3 g.)

Tomato Soup

$1\frac{1}{4}$ pints stock

$\frac{1}{2}$ pint milk

$\frac{1}{4}$ pint cream

12 ounces tomato puree

1 ounce brown sugar

2 tablespoons wine vinegar

1 teaspoon dried sage

 9 A group of friends have measured their heights and weights in imperial units.

Copy and complete the table to give each measurement in metric units to the nearest centimetre and kilogram.

	Height		Weight	
	Imperial	**Metric**	**Imperial**	**Metric**
Carl	6 ft	cm	11 st	kg
Hardeep	5 ft 11 in	cm	10 st 5 lb	kg
Kate	5 ft 5 in	cm	8 st 12 lb	kg
Viv	5 ft 3 in	cm	7 st 9 lb	kg

Note: ft = feet   in = inches   1 ft = 12 in   1 inch = 2.54 cm

st = stones   lb = pounds   1 st = 14 lb   1 kg = 2.2 lb

Homework 2

 1 a Estimate each of these lengths in centimetres:

  i 2 feet   **ii** 7 feet   **iii** $4\frac{1}{2}$ feet   **iv** 8 inches   **v** 11 inches

  b Estimate each of these lengths in metres:

  i 60 feet   **ii** 18 feet   **iii** 200 feet   **iv** $8\frac{1}{2}$ feet   **v** 50 inches

2 a Estimate each of these masses in pounds:

  i 4 kg   **ii** 1.5 kg   **iii** 3.25 kg   **iv** 500 g

  b Estimate each of these masses in kilograms:

  i 6 pounds   **ii** 56 pounds   **iii** $18\frac{1}{2}$ pounds   **iv** 175 pounds

3 Estimate each of these capacities in litres:

    a 6 pints   **b** 10 gallons   **c** 35 gallons   **d** $4\frac{1}{2}$ gallons

4 Get Real!

Judy lives about 100 miles from Hull. She wants to convert this distance to kilometres for her Belgian pen pal who is coming to visit. Roughly how far is it in kilometres?

5 Sam says that 10 litres is about the same as 45 gallons.

    a Explain why this is wrong.

    b Give the correct capacity to the nearest 0.1 gallons.

6 The hundredweight was an old imperial unit of weight equal to 112 lb.

Roughly how many kilograms is this?

7 Get Real!

Imagine you are on holiday in Germany.

You want to buy the items on the shopping list.

Re-write the list, giving approximate metric amounts to the nearest 0.1 kilogram or 0.1 litre.

Shopping list

10 lb potatoes   2 lb sausages

$\frac{1}{2}$ lb cheese $\frac{1}{4}$ lb mushrooms

4 pints milk $\frac{1}{2}$ pint cream

 **8 Get Real!**

The chart gives the distances in kilometres between some of the capital cities in Europe.

Amsterdam

650	Berlin							
197	764	Brussels						
1093	1696	941	Edinburgh					
2331	2869	3141	2795	Lisbon				
480	1074	333	608	2187	London			
1790	2364	1600	2254	651	1646	Madrid		
510	1051	320	1007	1821	399	1280	Paris	
1691	1502	1520	2404	2653	1796	2002	1389	Rome

Find the distance in miles, to the nearest 5 miles, for each of the following journeys:

a Brussels to Madrid

b London to Edinburgh

c Madrid to Paris

d Berlin to Rome

e Paris to Amsterdam

f Berlin to Lisbon

9 Get Real!

A drawing in an estate agent's leaflet gives the dimensions of the rooms in a studio flat in feet (ft).

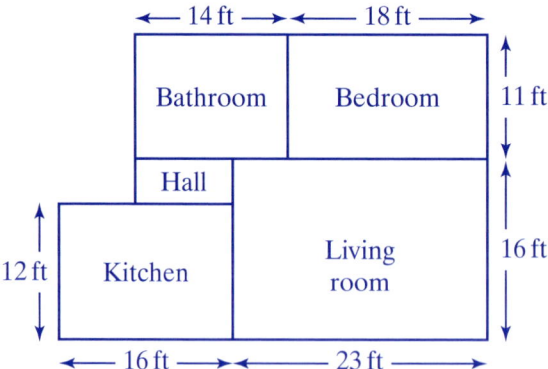

Draw a table that gives the approximate length and width of each room in metres.

10 **Get Real!**

A cookbook contains these recipes.

Write the ingredients in metric units, giving masses to the nearest 25 g and capacities to the nearest 10 mℓ.

Marmalade
3 lb Seville oranges
6 lb sugar
2 pints water

Orange cheesecake
12 oz cottage cheese
5 oz caster sugar
4 oz butter
6 oz digestive biscuits
1 oz cornflour
$\frac{1}{4}$ pint orange juice
2 eggs

oz means ounces and 1 lb = 16 oz

Homework 3

1 **a** Write down which metric unit of length you would use to measure each item listed below:

i length of a garden **iv** width of a road

ii thickness of a book **v** height of a tree

iii length of a lake **vi** length of the Isle of Man

b Write down which imperial unit of length you would use to measure each item above.

Choose from **inches**, **feet** or **miles**.

2 Which metric unit would you use to weigh each item listed below?

a luggage **c** bag of coal **e** jar of coffee

b packet of tea **d** lorry

3 Which metric unit would you use to measure the capacity of each container listed below?

a bottle of milk

b bottle of medicine

c bottle of salad dressing

d reservoir

4 Choose the most likely measurement of each item:

Length of a carrot 15 mm 15 m 15 cm	Height of a kitchen table 73 mm 73 cm 73 m	Width of an envelope 110 mm 110 cm 0.11 m
Length of a football pitch 10000 mm 1000 cm 100 m	Distance between motorway junctions 260 m 26000 mm 26 km	Diameter of a dinner plate 250 mm 25 mm 2.5 cm

5 Choose the most likely mass of each item:

Tub of margarine 50 g 500 g 5000 g	Bag of coal 20 g 20 kg 20 tonnes	Bull 12 000 g 1200 kg 120 t
Toddler 2000 mg 2000 g 20 kg	Envelope 40 g 4 g 40 mg	Banana 1.5 g 0.15 kg 0.0015 t

6 Choose the most likely capacity of each item:

Bucket 10 ml 10 cl 10 l	Wine glass 200 ml 200 cl 200 l	Teapot 15 ml 1.5 cl 1.5 l
Tube of hand cream 1.5 ml 15 ml 150 ml	Tropical fish tank 20 l 200 ml 20 cl	Dose of medicine 5 ml 50 cl 0.5 l

7 **a** Estimate the length, width and height of your school.

   b Estimate the length and width of the car park at your school.

8 Write down the names of five tourist attractions that are near where you live and estimate their distances from your school or college.

9 The following estimates are incorrect. Say why and in each case give a better estimate.

   a A hamster weighs 90 kg.

   b My arms are 45 mm long.

   c I use about 65 mℓ of water in my bath.

   d A bag of rice weighs about 2 mg.

   e A tablecloth measures 120 m by 75 m.

   f A watering can holds 10 mℓ.

10 **a** Name three things that you think could measure about 30 cm.

   b Name three items that you think could weigh about 200 g.

   c Name three containers that you think could hold about 500 mℓ.

11 The diagram shows a woman standing next to a wind turbine.

Estimate the height of the wind turbine.

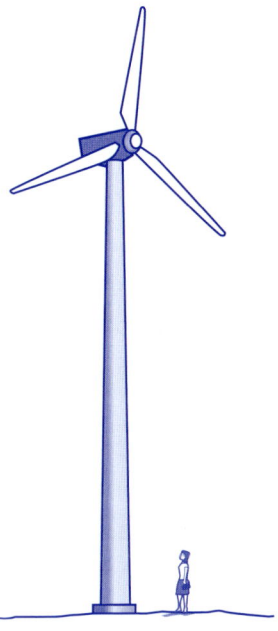

12 The diagram shows a dog at the side of an elephant.

   a Estimate the length of the elephant.

   b Estimate the height of the elephant.

Homework 4

 1 Find the average speed in each case:

   a A lion took 6 seconds to run 150 metres.

   b A swimmer swam 150 m in 5 minutes.

   c A train travelled 292 miles from Bristol to Newcastle in 5 hours.

 2 In each case find the distance travelled:

   a A lorry travels for 4 hours at an average speed of 58 km/h.

   b A spider runs at a steady speed of 4 cm/s for 7.5 seconds.

   c Viv skied for 15 minutes at a steady speed of 12 mph.

   d A walker travels for 2 hours at an average speed of 1.5 m/s.

 3 How long does each of these journeys take?

   a A bus completes a 240 km journey at an average speed of 60 km/h.

   b A dog runs for 120 metres at an average speed of 15 m/s.

   c A motorcyclist travels 90 miles at an average speed of 60 mph.

   d Biscuits move along a 10 m section of a production line at a steady speed of 2 cm/s.

4 Copy and complete the table.

Distance travelled	Time taken	Average speed
150 metres	25 seconds	
360 kilometres	5 hours	
	30 seconds	4.5 m/s
	$3\frac{1}{2}$ hours	64 mph
90 metres		15 m/s
12 metres		8 cm/s

 5 Get Real!

The table gives some of the results from women's athletics in the 2004 Olympic Games.

Winner	Distance	Time
Yuliya Nesterenko	100 m	10.93 s
Veronica Campbell	200 m	22.05 s
Tonique Williams-Darling	400 m	49.41 s
Kelly Holmes	800 m	1 min 56.38 s
Kelly Holmes	1500 m	3 min 57.90 s
Meseret Defar	5000 m	14 min 45.65 s
Xing Huina	10000 m	30 min 24.36 s

Find the average speed of each runner in:

a metres per second   **b** kilometres per hour.

 6 Five students are told that an insect flies 1.5 kilometres in half an hour.

The answers they give for the average speed are:

Kelly 45 km/h   Lee 1.2 m/s   Steve 50 m/s

Tom 20 cm/s   Meena 83 cm/s

a Who gave the correct answer?

b Explain what you think each of the other students did wrong.

 7 The average speed for a journey was 5 metres per second.

a Give a possible distance in metres and time in seconds.

b Give a possible distance in kilometres and time in minutes.

8 A motorcyclist starts a journey of 225 miles at 1 p.m.

  a He expects to travel at an average speed of 50 mph.
  What time does he expect to arrive?

  b If his motorcycle's rate of petrol consumption is 15 miles per litre,
  how much petrol will he use:

  i in litres   **ii** in gallons (to the nearest gallon).

9 Get Real!

  a i Find the distance you travel in a journey that you do regularly.

  ii Time how long this journey takes on a number of occasions.

  b Work out your average speed on each journey and your overall
  average speed.

10 An ambulance sets off at 10:35 a.m. to pick up a patient who lives 12
miles from the hospital. Its average speed on the journey is 40 mph.

  a At what time does the ambulance pick up the patient?

  The ambulance immediately returns to the hospital, arriving there at
  11:17 a.m.

  b What is its average speed:

  i on the return journey   **ii** for the whole journey?

11 Copy and complete the table.

Mass	Volume	Density
745 g	50 cm^3	
6.4 kg	0.8 m^3	
	24 cm^3	3.5 g/cm^3
	1.3 m^3	560 kg/m^3
342 g		0.9 g/cm^3
3.6 kg		450 kg/m^3

12 A concrete post weighs 3.6 tonnes. Its volume is 1.5 m^3.
Find its density in:

  a kg/m^3 **b** g/cm^3.

13 The density of butter is 0.86 g/cm^3.
Find the volume of butter in a 200 gram block.

14 The density of hydrochloric acid is 1200 kg/m^3.
What mass of acid would fill a 20 litre tank? (1 litre = 1000 cm^3)

15 The density of copper is 8.9 g/cm^3 and the density of zinc is 7.1 g/cm^3.

800 g of copper and 200 g of zinc are melted together to form an alloy.

Find:

a the volume of the copper   **c** the density of the alloy.

b the volume of the zinc

16 Find the value of the quantity named in each part:

a

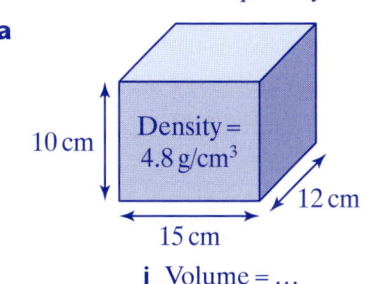

Density = 4.8 g/cm^3

10 cm

12 cm

15 cm

i Volume = ...

ii Mass = ...

b

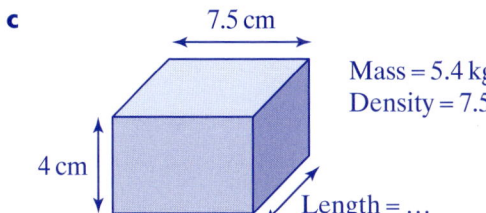

Mass = 1.8 kg

10 cm

10 cm

45 cm

i Volume = ...

ii Density in g/cm^3 = ...

c

7.5 cm

Mass = 5.4 kg

Density = 7.5 g/cm^3

4 cm

Length = ...

17 Population density is the number of people per square kilometre who live in an area.

a A city covers an area of 45 square kilometres and has a population of 89 415. What is its population density?

b A village covers 2.4 square kilometres and has a population density of 1545 people per square kilometre. How many people live in the village?

Homework 5

1 In each part, write down the upper and lower limits of the measurement.

a The height of a tree is 27 m to the nearest metre.

b The length of a road is 34 km to the nearest kilometre.

c The weight of a piece of cardboard is 15 g to the nearest gram.

d A boy's height is 163 cm to the nearest centimetre.

e The temperature in a room is 21°C to the nearest degree.

f The estimated length of a programme is 50 minutes to the nearest minute.

2 Give the upper and lower bounds of the following measurements.

  a Weight of a dog = 4.8 kg to 1 decimal place.

  b The time taken by a swimmer in a 50 m race was 52.46 seconds to 2 decimal places.

  c Temperature in an oven = 190°C to the nearest 10°C.

  d Amount of water in a tank = 4300 litres to the nearest 100 litres.

3 Give the range of possible values of the following measurements in the form $a \leqslant$ measurement $< b$.

  a The width of a river is 17 m to the nearest metre.

  b The length of a room is 4250 mm to the nearest 10 mm.

  c The weight of an elephant is 3200 kg to the nearest 100 kg.

  d The capacity of a coffee pot is 1.75 ℓ to 2 decimal places.

  e The temperature in a freezer is −18.0°C to the nearest tenth of a degree.

4 Get Real!

The table gives the heights of the five tallest mountains in the world to the nearest 5 metres.

Mountain	Height (metres)
Everest	8850 m
Qogir (K2)	8610 m
Kangchenjunga	8585 m
Lhotse	8500 m
Dhaulagiri	8200 m

Draw a table giving the lower and upper bounds for each height.

5 A plan says that the height of a building is 19 metres to the nearest metre.

Stella writes, 'If the height is x m then $18.5 \leqslant x < 19.5$'

James writes, 'If the height is x m then $18.5 \leqslant x < 19.4$'

Who is correct? Give a reason for your answer.

6 The length of a room is given as 4800 millimetres.

  a State whether the length has been given to the nearest mm, 10 mm or 100 mm if the least possible length is:

  i 4795 mm   **ii** 4799.5 mm   **iii** 4750 mm.

  b i How should 4800 mm be written if it is correct to the nearest tenth of a millimetre?

  ii In this case what is the least possible length?

7 What can you say about the accuracy of the length of the sides of an object if the length is given as:

  a 5 m   **b** 5.0 m   **c** 5.00 m?

Homework 1

1 £1 = 2.25 Australian dollars

£1 = 10 South African rand

a Using axes like those below, draw one conversion graph to convert pounds to Australian dollars and another to convert pounds to South African rand.

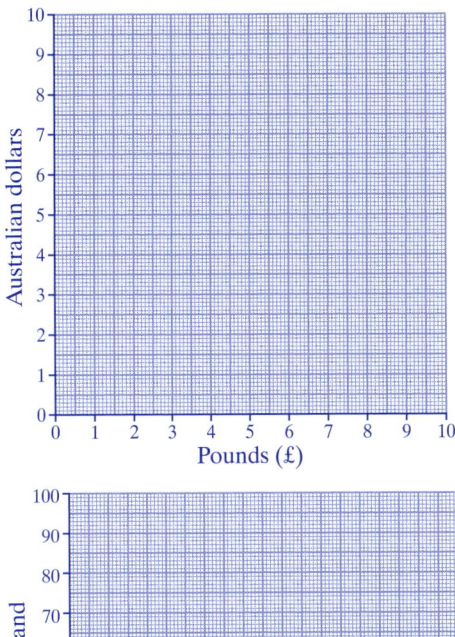

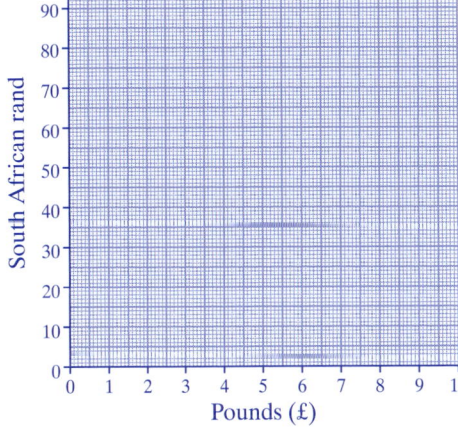

b Use your graphs to convert £7.50 to:

   i Australian dollars **ii** South African rand.

c Use your graphs to decide which is greater: 10 Australian dollars or 43 South African rand.

2 Mark is out walking. He descends a hill (height 225 m) at a constant rate of 20 m per minute.

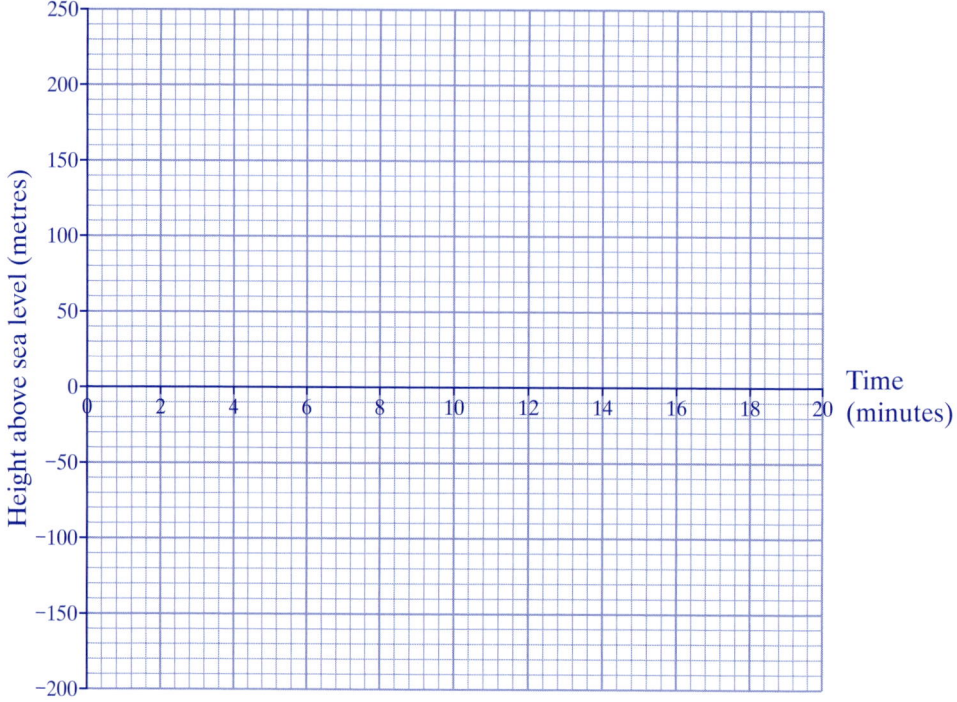

a Draw a graph to show how Mark's height above sea level varies with time.

Copy the set of axes below using each large square to represent 2 minutes on the horizontal axis and 50 metres on the vertical axis.

b Using your graph, find Mark's height above sea level after:

i 4 minutes **ii** $7\frac{1}{2}$ minutes **iii** 12 minutes **iv** 20 minutes.

c Using your graph, state when Mark is at sea level.

3 The following conversion graph can be used to convert kilograms (kg) to pounds (lb).

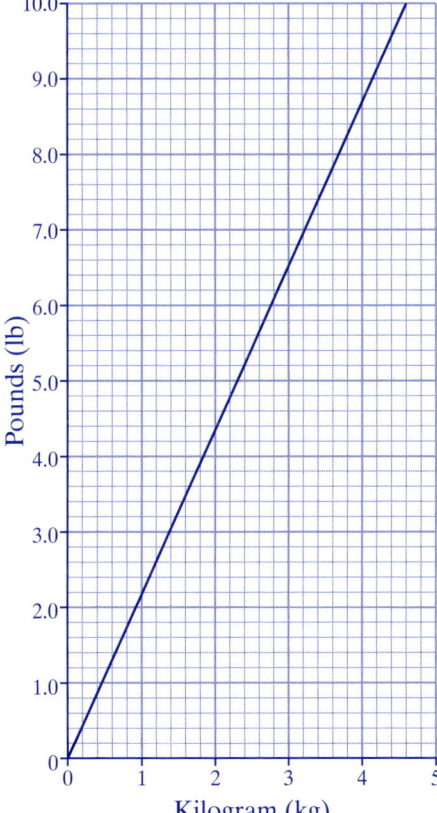

a Use the conversion graph to convert the following weights into pounds:

i 1.4 kg   **ii** 3.8 kg

b Use the conversion graph to convert the following weights into kilograms:

i 9.4 lb   **ii** 5.2 lb

Homework 2

1 The graph below shows the journeys of three people – Alf, Bob and Carl.

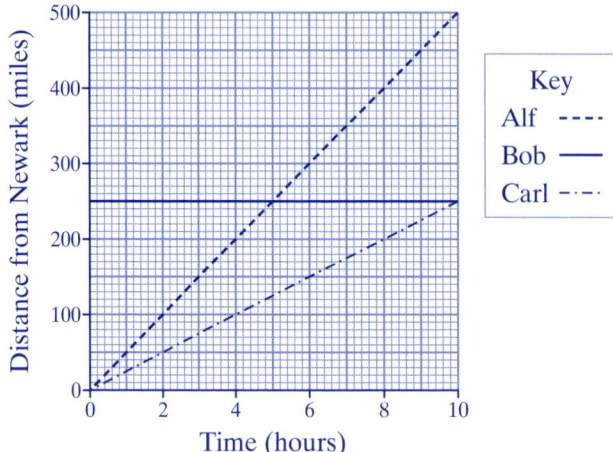

a Who does the graph suggest is travelling by:

   i car   **ii** scooter?

b Who was the furthest distance from Newark after:

   i 2 hours   **ii** 6 hours?

c How far is Bob from Newark after:

   i 2 hours   **ii** 4 hours?

d What can you say about Bob's journey?

e Who travelled the fastest?

f Calculate Carl's average speed in miles per hour (mph).

2 Mo the courier delivers meals-on-wheels from her depot in Yumville.
She leaves the depot at 10:45 a.m.

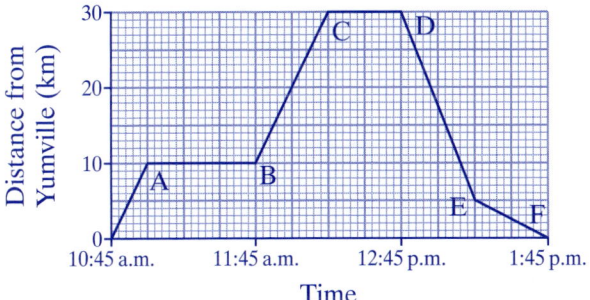

a How many times does Mo stop to deliver meals?

b How far has Mo travelled by 11 a.m.?

c Calculate Mo's speed during section OA in kilometres per hour (km/h).

d Calculate Mo's average speed over the first $1\frac{1}{2}$ hours of the journey.

e At what time does Mo arrive back at the depot?

f Calculate Mo's average speed for the whole journey.

3

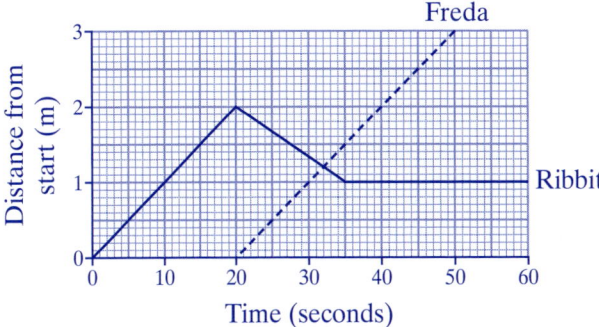

Write a commentary for the Three-metre Frog Race between Freda
and Ribbit.

You should include key distances, times and speeds in your commentary.

4 Luigi leaves his house at 9:30 a.m. and cycles to the post office, which
takes 20 minutes at 12 km/h. He queues for 40 minutes at the post office
and then cycles home at a speed of 16 km/h.

a Draw a distance–time graph representing Luigi's journey.

b At what time does Luigi arrive home?

9 Formulae

1   Write an expression for each of these statements.

   a Six times the number a.

   b Five times the number b plus 12.

   c The sum of four times the number c and twice the number d.

   d The product of the numbers e, f and g.

   e There are 30 students in each of x classes. What is the total number of students?

   f There are 11 footballers in a team. How many footballers would there be in y teams?

   g There are 7 girls in a netball team. How many would there be in n teams?

2   Write an expression for each of these statements.

   a x tickets at 60p each and y programmes at 45p each.

   b h sweets at 35p each and j biscuits at 15p each.

   c Andrew is 9 years old and his younger sister Claire is y years old. How much older is Andrew than Claire?

   d What is the total mass of 8 parcels at k kg each?

   e Abi gets £m pocket money. She spends £5. How much is left?

3   Write an expression for each of these statements. Let the unknown number be x.

   a Think of a number. Multiply by 6.

   b Think of a number. Multiply by 2. Subtract 3.

   c Think of a number. Square it. Add 2.

   d Think of a number. Double it. Add 5. Divide the result by 2.

4 Write an equation and solve it to find the unknown number in each of these questions. Let the unknown number be x.

   a Think of a number. Add 12. The answer is 17.

   b Think of a number. Multiply by 5. The answer is 80.

   c Think of a number. Multiply by 3. Subtract 1. The answer is 5.

   d Think of a number. Multiply by 9. Subtract 16. The answer is 11.

   e Think of a number. Multiply by 5. Add 2. The answer is 37.

5 Write down an equation for each of these diagrams. Solve it to find the value of x.

a 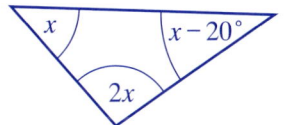   **b**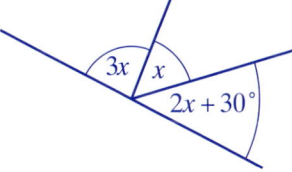

Not drawn accurately

6 The width of a rectangle is w cm. The length of the rectangle is 20 cm more than its width.

   a Write an expression for the length of the rectangle.

   The perimeter of the rectangle is 180 cm.

   b What are the lengths of the sides?

7 A square has sides of length $(2x - 10)$ cm. The perimeter is 40 cm. What is the area of the square?

8 The length of a rectangle is double its width. The perimeter is 36 cm. What is the area of the rectangle?

Homework 2

 1 Find the value of each of these expressions if $a = 1$, $b = 3$, $c = 5$ and $d = 2$

   a $3a$ **c** $a + b + d$ **e** $4d - 3a$ **g** $b^2 - 4d$

   b $4c$ **d** $b + 4a - c$ **f** $c^2 + b - a$ **h** $b^2 - a - c$

 2 If $e = 2$, $f = 4$ and $g = 0.5$, find the value of:

   a $4g$ **c** $f - g$ **e** $6g - 4f$ **g** efg

   b $f + 2e$ **d** $3e - g + 4f$ **f** $\dfrac{eg}{f}$ **h** $\dfrac{e^2 g}{3f}$

3 If $h = 4$, $l = -2$, $m = 3$ and $n = 1$, find the value of:

a ln **c** $6n - 3h$ **e** $mn - l$ **g** $m^2 - h^2 - l^2$

b $4h - 2m + l$ **d** $2h^2 + m + l$ **f** $\dfrac{h}{2} - l$ **h** $\dfrac{hmn}{l}$

4   Get Real!

By paying a deposit of £80 and then paying £50 a month for 10 months, a student can go on the school skiing trip.

a How much would have been paid after 4 months?

b Find a formula for the amount paid after n months.

c Substitute $n = 10$ into your formula to find the total cost of the holiday.

5 The formula for finding the distance travelled in kilometres (d) is $d = st$, where s is the speed in km/h and t is the time in hours.
What is the distance travelled when:

a $s = 70$ km/h and $t = 4$ h   **b** $s = 95$ km/h and $t = 1.5$ h

6 Lucia is finding the volume of a box using the formula $V = lbh$, where V represents the volume, l represents the length, b represents the width and h represents the height of the box.
What is the volume of the box when:

a $l = 10$ cm, $b = 8.5$ cm and $h = 3$ cm

b $l = 6$ cm, $b = 4.5$ cm and $h = 20$ cm

7 In science, Lauren is using the formula $F = ma$
Find F if:

a $m = 20$ and $a = 7.5$ **b** $m = 50$ and $a = 6.5$

8 In geography, Nick is using the formula $C = \dfrac{5(F - 32)}{9}$

Find C if:

a $F = 50$ **b** $F = -31$

9 In maths, Emily is using the formula $A = \pi r^2$ to find the area of two circles. Take $\pi = 3.14$ and work out the area of each circle:

a $r = 14.6$ mm   **b** $r = 4.5$ cm

10 In science, Lore is using the formula $R = \dfrac{PQ}{P+Q}$

Find R if:

a $P = 5$ and $Q = 15$ **b** $P = 10$ and $Q = 2$

11 In science, Sarah is using the formula $D = \dfrac{(U+V)T}{2}$

Find D when $U = 9$, $V = 61$ and $T = 7$

Homework 3

1 Rearrange each of these formulae to make x the subject:

a $a + x = b$ **c** $e = f + x$ **e** $x + 2k = l$ **g** $2x - 3q = r$ **i** $5 - x = u$

b $x - c = d$ **d** $g = x - h$ **f** $3m + 2x = p$ **h** $sx + 4 = t$

2 Rearrange each of these formulae to make y the subject:

a $qy - r^2 = s$ **c** $v^2 = w + xy$ **e** $\frac{1}{2}y - 3b = 4c$

b $t^2 + 4y = u^2$ **d** $zy = 4a$ **f** $d + y = e^2 - f$

3 Rearrange the formula $y = mx + c$ to make:

a c the subject   **b** m the subject.

4 Make the letter in brackets after each formula the subject:

a $GH = k$ (H) **c** $I = PRT$ (P) **e** $v = u + at$ (a)

b $E = mgh$ (h) **d** $ax + by = c$ (b) **f** $t = \dfrac{d}{s}$ (d)

5 These are all scientific formulae. Rearrange them so that the letter in brackets after each formula is the subject.

a $C = 2\pi x$ (x) **c** $V = \frac{1}{3}x^2 h$ (h) **e** $V = lbx$ (b)

b $V = x^2 h$ (h) **d** $e = Rx^2$ (R) **f** $P = Ri^2 t$ (t)

6 Rearrange the formula $v = Ri$ to make i the subject.

7 Rearrange the formula $s = vt$ to make v the subject.

8 Rearrange the formula $v = u + at$ to make:

    a u the subject   **b** a the subject   **c** t the subject.

9 Rearrange the formula $e = v + Ri$ to make:

    a v the subject   **b** R the subject   **c** i the subject.

10 Rearrange the formula $y = \dfrac{2x + 3}{5}$ to make x the subject.

11 Rearrange the formula $y = \dfrac{3x - 2}{2}$ to make x the subject.

10 Construction

1

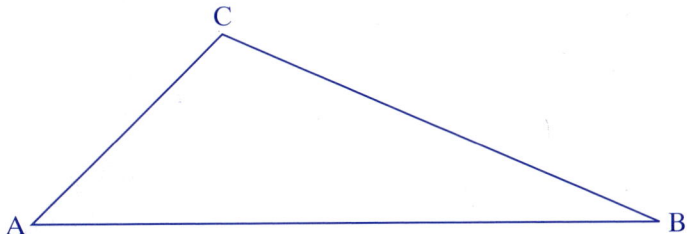

   a Measure and write down, to the nearest millimetre, the lengths of the sides of the triangle.

   b Measure accurately and write down the sizes of angles A, B and C.

   c Find the total of your three answers in part **b**.

2 Make an accurate drawing of parallelogram ABCD with AB = 4 cm, AD = 5 cm and BD = 4.5 cm.

   Measure and write down the length of AC on your drawing.

Not drawn accurately

3 Draw accurately triangle PQR with PQ = 5.8 cm, PR = 9.4 cm and angle P = 104°.

   What is the length of QR?

4 Get Real!

   A trawler and a yacht are 4 nautical miles apart with the yacht being due south of the trawler.

   A lighthouse is on a bearing of 132° from the trawler and on a bearing of 067° from the yacht.

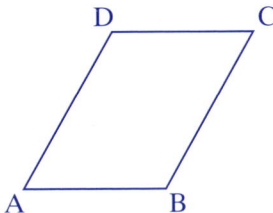

   Using a scale of 1 cm to represent 1 nautical mile, draw an accurate diagram showing the positions of the trawler, the yacht and the lighthouse.

   What is the distance between the yacht and the lighthouse?

5 An isosceles trapezium has base length 10.4 cm. Both diagonals measure 9 cm. The length of the sloping sides is 5.2 cm.

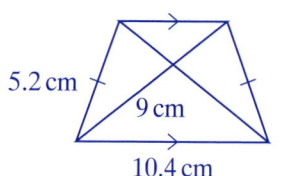

Not drawn accurately

5.2 cm

9 cm

10.4 cm

a Draw the trapezium accurately.

b Measure and write down the base angles.

c What is the width of the shorter parallel side?

Homework 2

1 **a** Use your protractor to draw an angle of 70°.

b Construct the angle bisector.

Check that each half-angle measures 35°.

2 **a** Draw a line of length 8 cm.

b Construct the perpendicular bisector of the line.

Check that each half measures 4 cm.

3 **Get Real!**

Hana would like to be selected to throw the discus in the next Olympics.

The discus arena is a V-shape with an angle of 40°.

Hana wants to keep her throws as central as possible.

Draw a diagram of the arena and construct the angle bisector so that Hana can keep a record of where her discus lands.

4 **Get Real!**

An artist is marking out a design based on the points of the compass on the floor of a hotel foyer.

Can you construct the same design using a ruler and compasses only?

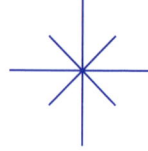

5 **a** Construct triangle PQR with PQ = 9 cm, PR = 5 cm and QR = 6 cm.

b Measure and write down the size of angle Q.

c Construct the bisector of angle Q, which cuts PR at X.

d Measure and write down the length of QX to the nearest millimetre.

Homework 3

1 **a** Draw a line of length 9 cm.

    b Mark a point above the line.

    c Construct the perpendicular from the point to the line.

2 **a** Draw a line of length 8 cm.

    b Using your line as the base, construct an isosceles triangle with base angles of 45°.

    c Measure and write down the length of the other two sides of your triangle.

3 **a** Using a ruler and compasses only, construct a square of side 5 cm.

    b Construct the angle bisector of one of the angles.

       If your constructions are accurate it should pass through the opposite vertex.

<u>4</u> Without using a protractor, construct an angle of 135°.

<u>5</u> Get Real!

A jeweller wants a diamond logo for his brochures.

Rachel designs this rhombus for him.

Construct an accurate full-size copy of the logo, using a ruler and compasses only.

What is the height of your logo to the nearest millimetre?

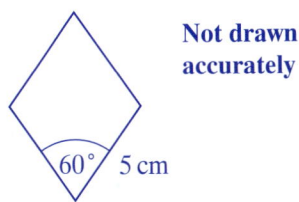

Not drawn accurately

60°   5 cm

Homework 4

1 Which of these are nets of a cuboid?

a

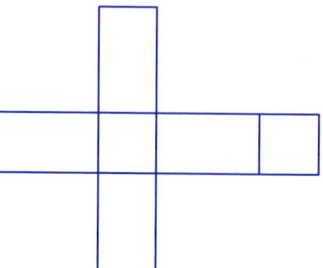

c

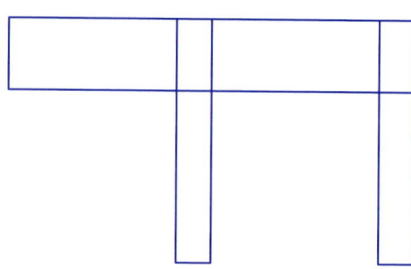

b

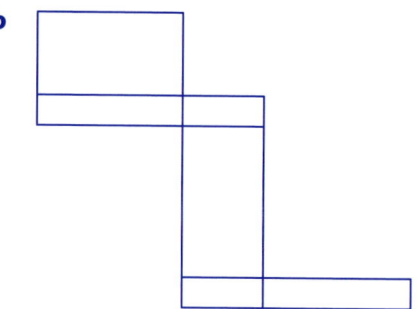

d

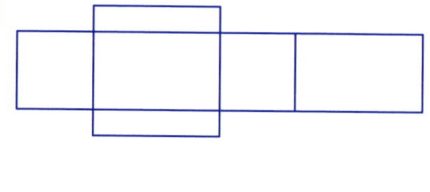

2 Which of these are nets of open cones, which are nets of open cylinders and which are neither?

a

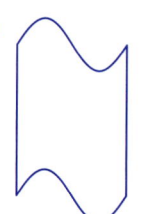

c

e

b

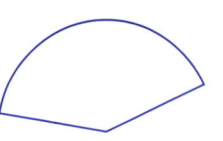

d

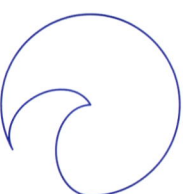

f

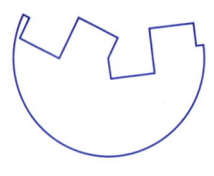

3 Get Real!

This is a picture of a box containing a sample tube of toothpaste.

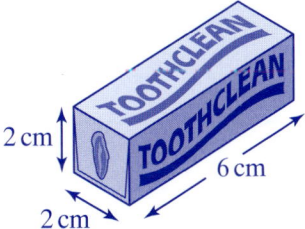

Draw an accurate net of the box using a ruler and protractor.

4 Get Real!

Olivia is designing a box to pack her homemade shortbread triangles ready to sell.

She sketches the box and adds on the measurements that she needs.

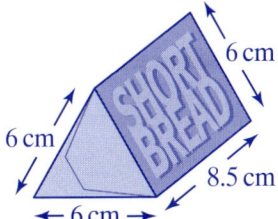

a Sketch a possible net for Olivia's shortbread box.

b Construct your net using a ruler and compasses only.

11 Probability

1   Five events are listed below.

   a Pigs will fly.

   b Your home will appear on TV.

   c You will have some maths homework this week.

   d There will be more people using the Internet next year than in 2000.

   e The next person to land on the moon will be female.

   Copy the probability scale and write the letters **a** to **e** at appropriate places.

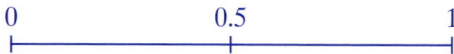

2   What is the probability of drawing a King from a pack of cards?

3   What is the probability of drawing a black 10 from a pack of cards?

4   300 raffle tickets are sold at the school fête.

   a Ian bought one ticket. What is the probability that he wins the raffle prize?

   b Rosie buys 5 tickets and her mother buys 10. What is the probability that one of them wins the raffle prize?

   c The probability that the prize will be won by someone in class 8Q is 0.1

     How many tickets were sold to class 8Q?

5   A letter is chosen from the word PREPOSTEROUS.

   a What is the probability that it is an A?

   b What is the probability that it is an S?

   c What is the probability that it is a P or an R?

   d What is the probability that it is a vowel?

6   Rajesh has a set of cards numbered from 11 to 40. He puts them face down on the table and asks Harry to pick a card.
What is the probability that Harry will pick:

a   a card numbered 5

b   a card numbered 21

c   a card with a number less than 15

d   a card with a number greater than 30

e   a card with a number with a 2 in it

f   a card with a number that is a multiple of 4?

7   Anna has 50 tiddlywinks in a box. Of these, 12 are red, some are white and the rest are blue. She picks a tiddlywink at random. The probability that she will pick a white tiddlywink is 0.5

a   How many white tiddlywinks are in the box?

b   What is the probability that she will pick a blue tiddlywink?

8   Jared has 20 marbles. Some are black, some are white and some are silver.

He picks a marble at random.

The probability that he will pick a black marble is 0.7

The probability that he will pick a white marble is 0.25

a   How many black marbles does he have?

b   How many white marbles does he have?

c   What is the probability that he will pick a silver marble?

Homework 2 ▨

1   A coin is tossed 150 times. The result is 79 heads and 71 tails. Todd says, 'You should get equal numbers of heads and tails, so there's something wrong with that coin'.

Is he correct? Give a reason for your answer.

2   A dice is thrown 300 times.

a   How many times would you expect to get an even number?

b   How many times would you expect to get a 5?

3 If the probability of getting an unwrapped toffee in a batch is 0.005, how many unwrapped toffees would you expect to find in a batch of 1000?

4 A spinner with nine equal divisions is spun 180 times. Three divisions are labelled X, two are Y, two are Z, one is P and one is Q.

a How many times would you expect the spinner to land on X?

b How many times would you expect it to land on Z?

c How many times would you expect it to land on Q?

5 The probability that an inhabitant of Random Island is left-handed is $\frac{1}{8}$ There are 4263 people living on Random Island. How many of them would you expect to be left-handed?

6 The table shows the frequency distribution after choosing a card at random from a full pack 520 times.

	Results from 520 random choices		
	Picture card (J, Q, K)	Ace	Number card
Frequency distribution	119	44	357

a What is the relative frequency of getting an Ace?

b Using theoretical probability, how many Aces would you expect?

c Which one of the results in the table is the closest to the result you would expect from theoretical probability?

7 There are 3 yellow discs, 4 green discs and 5 black discs in a box. A disc is chosen at random and then put back. This is done 120 times. The table shows the frequency distribution.

	Results from choosing a disc 120 times		
	Yellow	Green	Black
Frequency distribution	33	39	48

a What is the relative frequency of getting yellow?

b What is the relative frequency of getting yellow or green?

c What is the theoretical probability of getting black?

8 There are 50 counters in a bag. Some are blue, some are yellow and some are green. One counter is taken out at random and replaced. The frequency distribution table shows the results from 250 trials.

	Results from 250 trials		
	Blue	Yellow	Green
Frequency distribution	34	125	91

How many counters are there of each colour?

9 Bob carried out an experiment with a box of 30 numbered discs. He asked Ahmed and Kerry to take out a disc with their eyes shut, show it to him and then put it back. He recorded the results shown below.

	Results from 240 draws				
Number on disc	1	2	3	4	5
Frequency distribution	75	62	60	27	16

a Ahmed says, 'You must have had two discs with 5 on them'. Is Ahmed correct? Give a reason for your answer.

b Kerry says, 'You had the same number of discs with 2 and 3 on them'. Is Kerry correct? Give a reason for your answer.

Homework 3

1 **a** Draw a table to show all the possible outcomes when four coins are tossed.

Use your table to find:

b the probability of getting four heads

c the probability of getting one head and three tails

d the probability of getting two heads and two tails

e the probability of getting at least one tail.

2 Kirsty has six tiles and two cards as shown in the diagram.

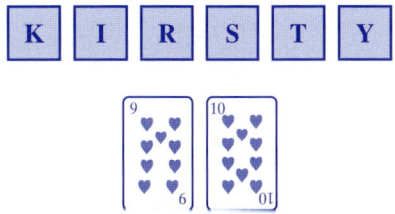

She takes one tile and one card at random.

a Draw a table of outcomes.

Use your table to find the probability that Kirsty takes:

b tile K and the 10 of hearts

c tile S or T and the 10 of hearts

d one of the tiles with a consonant and the 9 of hearts.

3 Two dice are thrown and the difference between their scores is recorded.

	1	2	3	4	5	6
1	0	1	2			
2						
3						
4						
5			1			
6			2			

a Copy and complete the sample space diagram.

Use your table to find:

b the probability of getting a difference of 4

c the probability of getting a difference of 6

d the probability of getting a difference greater than 4.

4 Gemma has the King, Queen and Jack of hearts plus the five tiles shown in the diagram.

She picks up one tile and one card at random.

a Draw a table of outcomes.

Use your table to find the probability that Gemma picks:

b the Jack and a tile with an E

c the Jack and a tile with an M.

5 Noah picks a card at random from a full pack. Harry picks a tile at random from a bag with five Hs and three Ys. Find the probability that:

a Noah picks a heart and Harry picks a Y

b Noah picks a black card and Harry's tile is an H

c Noah does not pick a club and Harry does not pick a Y.

Homework 4

1 Which of these are mutually exclusive events?

   a Throwing a two and an even number on a throw of a dice.

   b Throwing a number greater than 3 and an even number on a throw of a dice.

   c Picking a heart and a Queen from a pack of cards.

   d Picking the three of diamonds and the three of clubs from a pack of cards.

   e Picking a red ball and picking a blue ball from a bag containing red, white and blue balls.

2 The probability that Jason will be late for school tomorrow is 0.3

   What is the probability that he will not be late for school tomorrow?

3 The probability that Marie will fail her history exam is $\frac{3}{20}$
   What is the probability that she will pass the exam?

4 The probability that Ant will have chips for supper is 0.77

   What is the probability that he will not have chips?

5 The probability that AQA Rovers will win their next match is 0.45
   If the probability that they will draw is 0.25, what is the probability that they will lose?

6 Dilip can get into town in three different ways – with a lift from his father, by bus or walking. The probability that Dilip will get a lift is 0.65
   The probability that he will catch the bus is 0.3
   Baz says 'That means there is a probability of 0.32 that he will walk'.
   Explain why Baz is wrong and give the correct probability.

7 The table shows the probabilities that the arrow on a spinner will land on each number.

Colour	Probability
1	0.1
2	0.2
3	0.1
4	●
5	0.3

What is the number under the ink blot?

8 Restaurant Aqaluxe has a menu with four choices. The probability of a customer choosing each dish is shown.

Dish	Probability
Steak	0.57
Chicken	0.33
Monkfish	0.15
Vegetarian	0.05

a How can you tell that the probabilities are incorrect?

b The first probability is incorrect. What should it be?

9 Vicky goes out to buy some new jeans. The probability that she buys them from Jeans'R'Us is $\frac{1}{4}$
The probability that she buys them from the Superstore is $\frac{5}{12}$
What is the probability that she buys them from one of these stores?

12 Graphs of linear functions

Homework 1

1 Draw an x-axis and a y-axis, each going from -6 to 6.

   a On the axes, draw the lines $x = 5$ and $x = -3$

   b On the axes, draw the lines $y = -5$ and $y = 4$

   c Where do $x = 5$ and $y = 4$ meet?

   d Where do $y = 4$ and $x = -3$ meet?

2   **a** Copy and complete this table of values for $y = 3x$

x	−3	−2	−1	0	1	2	3
y		−6			3		

   b Draw an x-axis labelled from -3 to 3 and a y-axis labelled from -9 to 9.
     Plot the points in the table and join them with a straight line.

   c What are the coordinates of the point on the line where $y = 3$?

   d What are the coordinates of the point on the line where $x = -2$?

3   **a** Copy and complete this table of values for $y = x - 2$

x	−3	−2	−1	0	1	2	3
y		−4			−1		

   b Draw an x-axis labelled from -3 to 3 and a y-axis labelled from -5 to 5.
     Plot the points in the table and join them with a straight line.

   c Where does the line $y = x - 2$ cut:

     i   the x-axis   **ii**   the y-axis   **iii**   the line $x = 2$ **iv**   the line $y = -1$?

   d What are the coordinates of the point on the line where $x = -2$?

4 **a** Copy and complete this table of x-values and y-values for the graph
with equation $y = 3x - 3$

x	-2	-1	0	1	2	3
y	-9			0		

b The x-coordinates go up one unit each time. How many units do the
y-coordinates go up each time?

c Plot the points on graph paper and join them to make the straight-line graph
$y = 3x - 3$

d Find the y-coordinate of a point on the graph with an x-coordinate of 1.5

e Find the x-coordinate of a point on the graph with a y-coordinate of 2.

5 **a** For the graph of $x + y = 5$, complete these coordinate pairs:

$(-4, \_\_) \ (0, \_\_) \ (4, \_\_)$

This means that the x-coordinates and
the y-coordinates add up to 5

b Draw an x-axis and a y-axis, each going from -10 to 10, then plot the three
points and join them with a straight line.

c Write down the coordinates of the points where the graph $x + y = 5$ crosses
the x-axis and the y-axis.

d Use the equation of the graph to work out the coordinates of the points
where the graph crosses the x-axis and the y-axis. Show how you worked
it out.

6 Draw an x-axis and a y-axis, each going from -10 to 10.

a On the axes, draw these lines:

i $y = 4x$ **ii** $y = 2x - 1$

b i Where do the lines $y = 4x$ and $y = 2x - 1$ cut the x-axis and the y-axis?

ii How can you find this from the equations?

c Where do the lines $y = 4x$ and $y = 2x - 1$ meet?

7 Which of these points lie on the graph of $y = -\frac{1}{2}x$?

$(2, -1), (2, 0), (0, 0), (-2, -1), (1, -2), (2, -4), (0.6, 0.3)$

Show how you found your answers.

8 a Copy and complete these coordinate points for the line $y = 2x - 2$

(−2, __), (0, __), (__, 0)

b Draw an x-axis and a y-axis, each going from −6 to 6, then plot the points and join them with a straight line.

c Where does the line $y = 2x - 2$ cut:

i the x-axis **ii** the y-axis **iii** the line $x = 2$ **iv** the line $y = 2$?

9 a Find at least three points on the line $y = 2x - 3$

b Draw an x-axis and a y-axis, each going from −10 to 10. Plot the points and join them with a straight line.

c On the same diagram, draw the graph $y = 2x - 1$

d What is the same and what is different about the two graphs?

10 a Draw an x-axis and a y-axis, each going from −6 to 6. Draw the graphs $y = -x + 4$ and $y = -x - 4$

b What is the same and what is different about the two graphs?

c Explain how you can use the equations to tell you what is the same and what is different about the graphs.

11 The diagram shows several straight-line graphs.

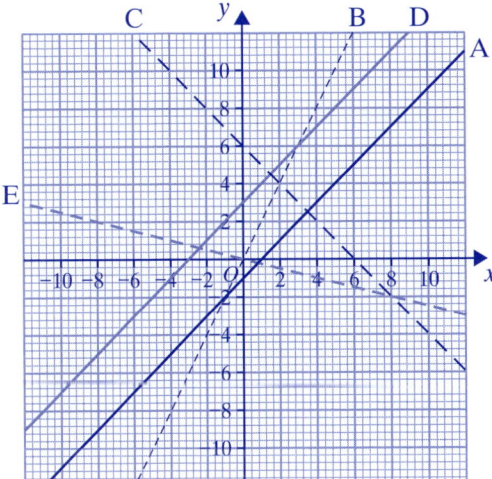

a Write down the letters of two graphs that are parallel.

b Write down the letters of two graphs that are perpendicular.

c Write down the letters of the graphs that go through the origin.

d Write down the letter of the steepest line.

e i Write down the coordinates of three points on graph A.

ii What is the equation of graph A?

Homework 2 ⊠

1 Which of these equations represent straight-line graphs?

    a $x + y = 5$ **b** $xy = 5$ **c** $y = \dfrac{x}{5}$ **d** $y = \dfrac{5}{x}$ **e** $y = x^2 + 5$

2 Rearrange each of these equations into the form $y = mx + c$

    a $x = 2y$ **b** $x + 2y = 0$ **c** $x + 2y = 3$ **d** $x - 2y = 0$

3 **a** Copy and complete the table of x-values and y-values for the graphs with equations $y = 3x$ and $y = 3x - 4$

x	-2	-1	0	1	2	3
$y = 3x$			0			
$y = 3x - 4$			-4			

    b Draw both graphs on the same diagram.

    c Use your diagram to find:

        i the gradient of each graph

        ii the point where each graph cuts the y-axis.

4 Repeat question **3** for the graphs $y = \frac{1}{2}x$ and $y = \frac{1}{2}x + 3$

5 Find the gradients of the straight-line graphs with these equations.

a $y = 3x$	**e** $y = 3x + 5$	**i** $y = \frac{1}{5}x$	**m** $y = -0.3x$
b $y = \frac{1}{4}x$	**f** $y = \frac{1}{4}x + 5$	**j** $y = -\frac{1}{5}x$	**n** $y = 0.3x - 0.2$
c $y = -3x$	**g** $y = -3x + 5$	**k** $y = \frac{1}{4}x - \frac{4}{5}$	**o** $y = 0.3x$
d $y = -\frac{1}{4}x$	**h** $y = -\frac{1}{4}x + 5$	**l** $y = -\frac{1}{5}x - \frac{4}{5}$	**p** $y = -0.3x - 0.2$

<u>6</u> Find the coordinates of the points where the lines in question **5** meet the y-axis.

7 a Write down the gradients and y-intercepts of the graphs in this diagram.

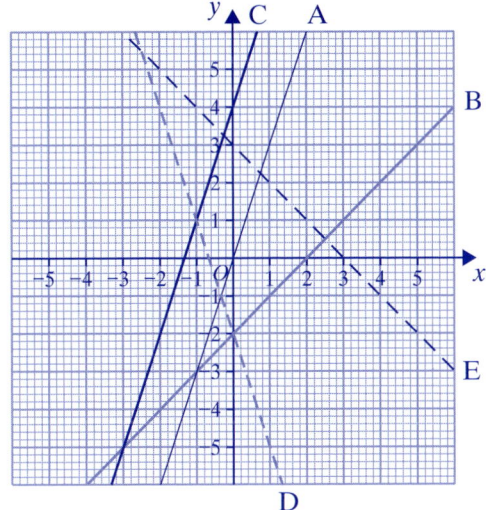

The y-intercept is the y-coordinate of the point where the line cuts the y-axis.

b Write down the equation of each graph.

8 a Rearrange each of these equations into the form $y = mx + c$

   i $y + x = 15$ **ii** $y - x = 15$ **iii** $y + x = -15$ **iv** $y - x = -15$

b Find the gradients of the graphs represented by these equations.

9 Rearrange each of these equations into the form $y = mx + c$ and find the gradients and y-intercepts of the corresponding straight-line graphs.

   a $3y + x = 9$ **c** $3y - x = -9$ **e** $3x + 5y = 15$ **g** $-3x + 5y = -15$

   b $y - 3x = 9$ **d** $y + 3x = -9$ **f** $3x - 5y = 15$ **h** $3x + 5y + 15 = 0$

10 Get Real!

The equation $y = \frac{8}{5}x$ gives a straight-line graph that converts distances in miles (x) to distances in kilometres (y).

a What is the gradient of the graph?

b What is its y-intercept?

c Interpret the meanings of these numbers.

d Copy and complete this table of values for the graph.

Distance (miles)	0	10	
Distance (kilometres)			40

e Draw the graph, choosing scales so that the three points in the table fit on the diagram.

f Use the graph to convert 44 miles to kilometres.

g What graph would convert distances in kilometres to distances in miles?

1 Draw a sketch and calculate the length of the hypotenuse of each of the following triangles.

Give your answer to the nearest millimetre.

a AB = 14 cm, BC = 22 cm, angle ABC = 90°

b XY = 3.9 cm, YZ = 2.3 cm, angle XYZ = 90°

c PQ = 27.4 cm, PR = 12.9 cm, angle RPQ = 90°

d DE = EF = 45 cm, angle DEF = 90°

e RS = 2ST, ST = 13 cm, angle RST = 90°

2 Get Real!

A gate is 2 metres wide.

The horizontal bars are 17 cm apart and each of the six bars is 2.5 cm thick.

Find the length of the diagonal (ignore its width).

Leave your answer as a square root.

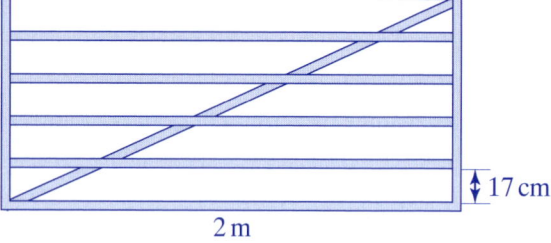

17 cm

2 m

3 Get Real!

The country of Sylvania has declared itself a no fly zone and will no longer allow aircraft to fly through its airspace.

Captain Walkington has had to submit a new flight plan for his journey.

He will fly 300 km south, then turn and fly 250 km east.

How many extra kilometres will he fly using this flight plan instead of flying direct?

4 Get Real!

Tru Grit is a company that designs containers to hold the grit used for gritting roads in icy conditions. The smallest grit container that the company makes is a prism. The cross-section of the prism is a trapezium.

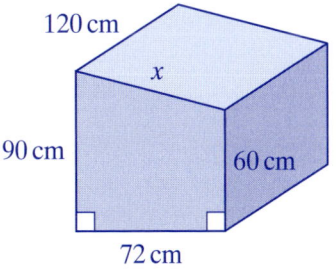

120 cm

x

90 cm

60 cm

Not drawn accurately

72 cm

a Find *x*.

b Work out the area of plastic used to make the grit container.

Homework 2

1 Draw a sketch and find the length of the missing side of each triangle.

Give your answer to an appropriate degree of accuracy.

a AB = 20 cm, AC = 12 cm, angle ACB = 90°

b XY = 20 cm, YZ = 29 cm, angle YXZ = 90°

c DE = 20 cm, EF = 15 cm, angle DEF = 90°

d PR = 20 cm, RQ = 48 cm, angle PRQ = 90°

e JK = 20 cm, KL = 20.5 cm, angle KJL = 90°

f VW = VX = 20 cm, angle WVX = 90°

g ST = 20 cm, US = UT, angle SUT = 90°

2 Get Real!

The diagonal length of a computer monitor's flat screen is 43 cm.

The screen's width is 34 cm.

Find the height of the screen.

Give your answer to the nearest millimetre.

3

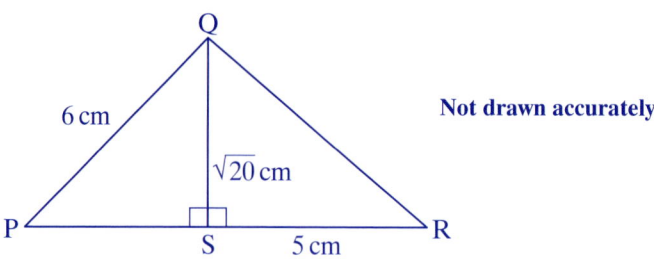

Not drawn accurately

PQ = 6 cm, QS = √20 cm, SR = 5 cm

Angles PSQ and RSQ are right angles.

a Find QR, leaving your answer as a square root.

b Find PS.

c Is triangle PQR right-angled? Give a reason for your answer.

4 Jing draws a triangle for each of these sets of measurements.

a 5 cm, 12 cm, 15 cm

b 8 cm, 15 cm, 17 cm

c 10 cm, 24 cm, 26 cm

d 2√3 cm, 6 cm, 18√2 cm

e 4√5 cm, 10 cm, 6√5 cm

Which of the triangles will be right-angled?
Give a reason for your answer.

5 Joseph calculates the area of this equilateral triangle as follows.

Using Pythagoras' theorem in triangle AXC:

$AC^2 = AX^2 + XC^2$

$8^2 = AX^2 + 4^2$

$AX = 48$ cm

Area of ABC $= \frac{1}{2} \times 8 \times 48 = 192$ cm^2

Is Joseph correct? Give a reason for your answer.

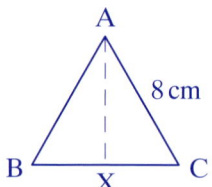

<u>6</u> Get Real!

The diagram shows the side view of the roof above a front door.

The vertical support is 108 centimetres.

The middle support is 117 centimetres and meets the middle of the roof.

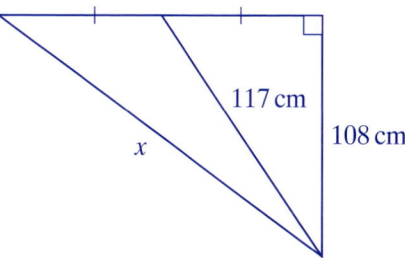

Find the value of x, to the nearest millimetre.

14 Quadratic graphs

 1 The graph shows the function $-x^2$ for values of x from -4 to 4.

Graph of $y = -x^2$

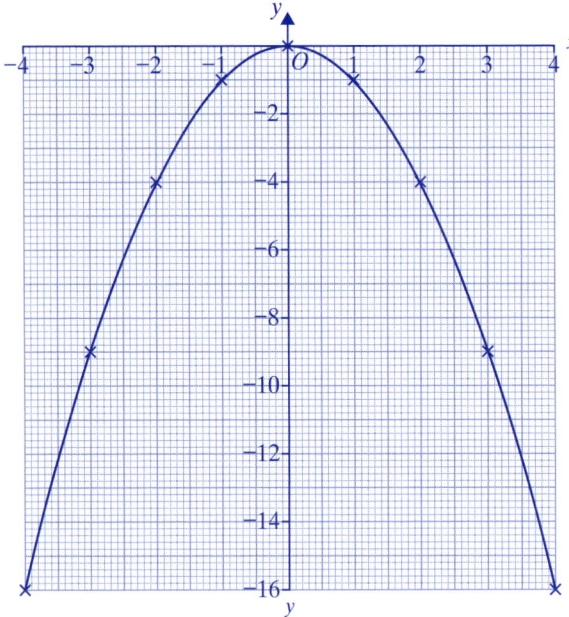

a Use the graph to estimate the value of:

    i -3.6^2 **ii** $-(-2.4)^2$

b Use the graph to find the values of x when:

    i $y = -11$ **ii** $y = -7.8$

 2 **a** Copy and complete this table for $y = x^2 - 3$

x	−3	−2	−1	0	1	2	3
y		1	−2				6

b Draw the graph of $y = x^2 - 3$ for values of x from -3 to 3.

c Use your graph to find the value of y when:

    i $x = 1.6$ **ii** $x = -2.3$

d Use your graph to find the values of x when:

    i $y = 4$ **ii** $y = -1.4$

3 **a** Copy and complete this table for $y = 4x^2$

x	−3	−2	−1	0	1	2	3
y	36				4		

b Draw the graph of $y = 4x^2$ for values of x from −3 to 3.

c Use your graph to find the value of y when:

    i $x = 2.4$ **ii** $x = -1.3$

d Use your graph to find the values of x when:

    i $y = 6$ **ii** $y = 27$

 4 **a** Copy and complete this table for $y = \dfrac{x^2}{3}$ and $y = \dfrac{2x^2}{3}$

x	−9	−6	−3	0	3	6	9
$y = x^2$	81		9				81
$y = \dfrac{x^2}{3}$	27		3				27
$y = \dfrac{2x^2}{3}$	54		6				54

b On the same axes of x and y draw the graphs of $y = \dfrac{x^2}{3}$ and $y = \dfrac{2x^2}{3}$

c Describe the similarities and differences between the graphs.

 5 **a** Copy and complete this table for $y = x^2 + 1$

x	−4	−3	−2	−1	0	1	2	3	4
y	17	10				2	5		

b Draw the graph of $y = x^2 + 1$ for values of x from −4 to 4.

c Give the y-coordinate of the point on the curve with an x-coordinate of:

    i   3.5   **ii**   −2.4

d Give the x-coordinate of the points on the curve with a y-coordinate of:

    i   14   **ii**   7

e Write down the coordinates of the lowest point on the curve.

6 **a** Copy and complete this table for $y = 16 - x^2$

x	−4	−3	−2	−1	0	1	2	3	4
x^2		9	4			1			16
$y = 16 - x^2$		7	12			15			0

b Draw the graph of $y = 16 - x^2$ for values of x from −4 to 4.

c Write down the coordinates of the points where the curve crosses the x-axis.

d Write down the coordinates of the highest point on the curve.

 7 **a** Copy and complete the table.

x	−4	−3	−2	−1	0	1	2	3	4
$2x^2$	32	18		2			8		
$y = 2x^2 - 5$	27	13		−3			3		

b Draw the graph of $y = 2x^2 - 5$ for values of x from −4 to 4.

c Use your graph to find the value of y when:

  i $x = 1.3$ **ii** $x = -3.2$

d Use your graph to find the values of x when:

  i $y = 20$ **ii** $y = 9$

 8 **a** Copy and complete the table below, then use it to draw the graph of $y = (x + 1)(4 - x)$

x	−2	−1	0	1	1.5	2	3	4	5
$x + 1$	−1				2.5		4		6
$4 - x$	6				2.5		1		−1
$y = (x + 1)(4 - x)$	−6				6.25		4		−6

b Write down the coordinates of the points where the curve crosses the x-axis.

 9 **a** Draw a table and a graph for $y = x(x + 2)$ for values of x from −4 to 2.

b Write down the coordinates of the points where the curve crosses the x-axis.

10 The graph shows the points that Cliff has plotted for his graph of $y = 2x^2$

Graph of $y = 2x^2$

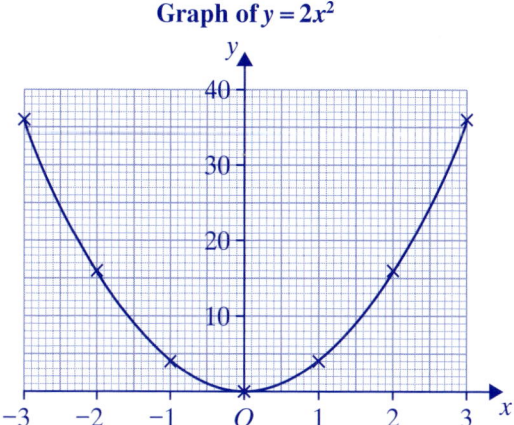

a Is Cliff correct?

b Give a reason for your answer.

11 The graphs of three quadratic functions are shown in the sketch.

The functions are

$$y = x^2 + 5 \qquad y = 5x^2 \qquad y = -5x^2$$

Choose the function that represents each curve.

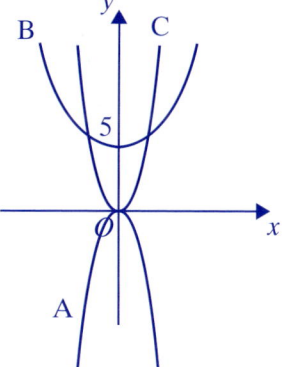

12 **Get Real!**

When a stone is dropped from a high bridge over a river, the height of the stone above the water, h metres, after t seconds is given by the formula $h = 76 - 5t^2$

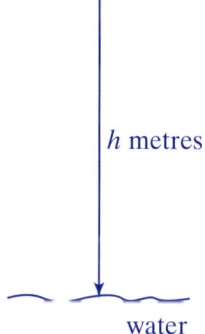

h metres

water

a Copy and complete the following table to give values of h for values of t from 0 to 4.

t (seconds)	0	1	2	3	4
t^2				9	
$5t^2$				45	
$h = 76 - 5t^2$ (metres)				31	

b Draw the graph of $h = 76 - 5t^2$ for $0 \leqslant t \leqslant 4$

c Use your graph to find:

   i the height of the stone after 1.5 seconds

   ii the time when the stone is 20 metres above the water

   iii the time when the stone hits the water.

Homework 2

 1   **a** Copy and complete this table for $y = x^2 + 2x$

x	−4	−3	−2	−1	0	1	2
y	8	3		−1		3	

    b Draw the graph of $y = x^2 + 2x$ for values of x from −4 to 2.

    c Use your graph to solve the equation $x^2 + 2x = 0$

    d i   Draw the line $y = 7$ on your graph.

       ii   Find the x-coordinates of the points where the line $y = 7$ crosses the curve $y = x^2 + 2x$

      iii   Write down the quadratic equation whose solutions are the answers to part **ii**.

 2   **a** Copy and complete this table for $y = x^2 - 2x + 1$

x	−2	−1	0	1	2	3	4
y	9				1	4	

    b Draw the graph of $y = x^2 - 2x + 1$ for values of x from −2 to 4.

    c i   Write down the x-coordinate of the point where the curve meets the x-axis.

      ii   Write down the quadratic equation whose solution is the answer to part **i**.

    d i   Draw the line $y = 5$ on your graph.

      ii   Write down the coordinates of the points where the line $y = 5$ crosses the graph of $y = x^2 - 2x + 1$

 3   **a** Copy and complete this table for $y = x^2 - x$

x	−2	−1	0	1	2	3
y	6				2	

    b Also calculate the value of y when $x = 0.5$

    c Use your answers to parts **a** and **b** to draw the graph of $y = x^2 - x$

    d Use your graph to solve the equation $x^2 - x = 0$

    e i   Draw the line $y = 4$ on your graph.

      ii   Find the x-coordinates of the points where the line $y = 4$ crosses the curve $y = x^2 - 3x$

      iii   Write down the quadratic equation whose solutions are the answers to part **ii**.

4 **a** Copy and complete this table for $y = 10 - x - x^2$

x	−4	−3	−2	−1	−0.5	0	1	2	3
y		4			10.25			4	

b Draw the graph of $y = 10 - x - x^2$ for $-4 \leqslant x \leqslant 3$

c i Write down the x-values where the curve crosses the x-axis.

  ii Write down the quadratic equation whose solutions are the answers to part **i**.

d i Write down the x-values where the curve meets the line $y = 6$

  ii Write down the quadratic equation whose solutions are the answers to part **i**.

5 **a** Copy and complete this table for $y = 2x^2 + x - 7$

x	−4	−3	−2	−1	0	1	2	3	4
y	21	8		−6			3	14	

b Calculate the value of y when $x = -0.25$

c Draw the graph of $y = 2x^2 + x - 7$ for values of x from −4 to 4.

d Find the coordinates of the points where the graph crosses:

  i the line $y = 0$ **ii** the line $y = 15$ **iii** the line $y = -7$

6 **Get Real!**

When a car does an emergency stop, the distance it travels in coming to a halt is related to its speed by the formula:

$d = 0.015v^2 + 0.3v$ where d is the distance in metres
                           and v is the speed in miles per hour.

a Copy and complete the following table.

v (mph)	0	10	20	30	40	50	60	70
d (m)			12		36	52.5		94.5

b Draw a graph of d against v using 2 cm to represent 10 mph on the v-axis and 10 m on the d-axis.

c Use your graph to estimate the distance it would take to stop if the car was travelling at:

  i 25 mph   **iii** 53 mph

  ii 35 mph   **iv** 68 mph.

d Use your graph to estimate how fast a car can travel and be able to stop in:

  i 10 m   **ii** 25 m   **iii** 75 m.

7 A teacher asks her class to draw up a table for $y = 2 - x - x^2$

    a This is Kylie's table.

x	-3	-2	-1	0	1	2	3
2	2	2	2	2	2	2	2
$-x$	$+3$	$+2$	$+1$	-0	-1	-2	-3
$-x^2$	$+9$	$+4$	$+4$	-0	$+1$	$+4$	$+9$
$y = 2 - x - x^2$	14	8	7	2	2	4	8

      i Are any of Kylie's y-values correct?

      ii Explain any mistakes she has made.

    b This is Will's table.

x	-3	-2	-1	0	1	2	3
2	2	2	2	2	2	2	2
$-x$	-3	-2	-1	-0	-1	-2	-3
$-x^2$	-9	-4	-1	-0	-1	-4	-9
$y = 2 - x - x^2$	-10	-4	0	2	0	-4	-10

      i Are any of Will's y-values correct?

      ii Explain any mistakes he has made.

 8   **a** Draw a table and graph for $y = 5x^2 + 4x - 8$ for $-3 \leqslant x \leqslant 2$

    b Use your graph to solve the following equations.

      i $5x^2 + 4x - 8 = 0$ **ii** $5x^2 + 4x - 8 = 10$

    c Emma says that the equation $5x^2 + 4x - 8 = -10$ cannot be solved.

      i Is she correct?   **ii** Give a reason for your answer.

 9   **a** Draw the graph of $y = 9 + 5x - 3x^2$ for $-2 \leqslant x \leqslant 4$

    b i Write down the solutions of the equation $9 + 5x - 3x^2 = 0$

      ii Explain how you found the solutions and why your method works.

15 Loci

1 Explore these loci practically.

 a Find the path of a corner of an equilateral triangle as it is rotated against a ruler.

 Draw the locus.

 b Find the path of a corner of an isosceles triangle as it is rotated against a ruler.

 Draw the locus.

2 **Get Real!**

 a Draw a sketch of the locus of the girl as she goes up and down on the seesaw.

 b How would the locus of the boy be different?

3 AB is perpendicular to BC. Draw the locus of points that are twice as far from BC as they are from AB.

 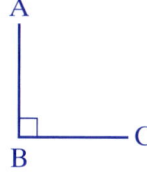

 > **HINT** Imagine AB is the *y*-axis of a coordinate grid, and BC is the *x*-axis. Can you find the coordinates of some points twice as far from the *x*-axis as from the *y*-axis?

4 Mark two points, A and B, 8 cm apart.

 Find two points that are 3 cm from A and 7 cm from B.

 Find two points that are 4 cm from A and 6 cm from B.

 Find some more points where the total distance from A and B is 10 cm.

 Join them up to draw the locus of all points with a total distance of 10 cm from A and B.

<u>5</u> Get Real!

Match the loci below with the descriptions that follow.

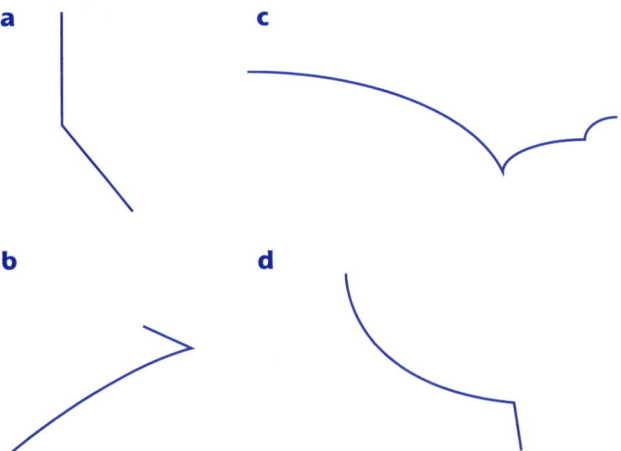

i   A football hitting a wall

ii A boy going down a slide and falling off the bottom

iii A man jumping out of an aeroplane and opening a parachute

iv A cricket ball being bowled, bouncing once and then being hit

Homework 2

1 Get Real!

A garden is 12 m long and 8 m wide.

a Using a scale of 1 cm to 2 m, make a scale drawing of the garden.

b The owner builds a patio, 2 m wide, along the edge AD. Draw this on your plan.

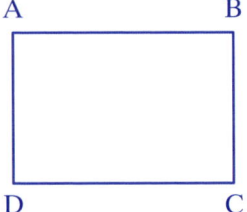

2 Get Real!

The next-door neighbour has a garden exactly the same size.

He plants a tree exactly in the middle of the garden.

a Construct a plan of the garden and mark the position of the tree.

b He grows a lawn in the garden, but he does not want grass within 2 m of the tree. He also wants a flowerbed, 1 m wide, along edges BC and CD. Shade in the area where he wants the lawn to be.

3 **a** Construct a right angled triangle ABC, where angle A = 90°, AB = 8 cm and AC = 11 cm.

    b Mark all the points that are the same distance from AB and BC, and closer to B than C.

4 Get Real!

A, B and C are three television transmitters.

A is 30 km due west of B, and C is 24 km due south of B.

    A   B
    ×   ×

    a Use a scale of 1 cm to 4 km to make a scale drawing of the transmitters.

                                        × C

    b Anyone within 12 km of transmitter B receives their signal from B. Beyond this, they use either transmitter A or C, whichever is the closest.

       Construct the area where transmitter A is used.

5 Get Real!

A man devises a brilliant plan to keep his two goats apart, which will still allow them to graze over the widest area of his field, which measures 22 m wide and 30 m long.

He erects two posts, 12 m apart, with a ring on the top of each post.

He passes a 23 m length of rope through the rings, and ties a goat to each end of the rope.

The diagram shows the position of the two posts in the field.

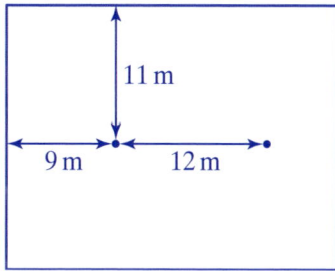

Make a scale drawing of the field, and shade the area the goats can reach.

16 3-D solids

Homework 1

1 a In the cube shown in the diagram, what shape is:

    i face PQUT

    ii face QRVU?

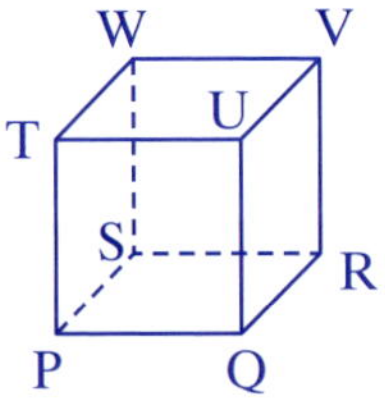

b i Write down the size of the angles of triangle PQT.

    ii Name a triangle that is congruent to triangle WPS.

    iii Name a pair of congruent right-angled triangles that are not isosceles.

2 a Sketch a cuboid that is 3 cm long, 2 cm wide and 1 cm high.

b Calculate:

    i the volume   **ii** the surface area of the cuboid.

c Repeat parts **a** and **b** for a cuboid that is 12 cm long, 8 cm wide and 4 cm high.

3 a Draw an isometric diagram of a cube with sides of length 5 cm.

Give the dimensions of the cube on your diagram.

b Calculate:

    i the volume   **ii** the surface area of the cube.

4 Draw:

a an isometric diagram

b a net

of a cuboid that is 6 cm long, 4 cm wide and 3 cm high.

Show the dimensions of the cuboid on each diagram.

5 The nets of two cuboids are drawn below.

Draw a diagram of each cuboid on dotty isometric paper.

Label each diagram to show the dimensions of the cuboid.

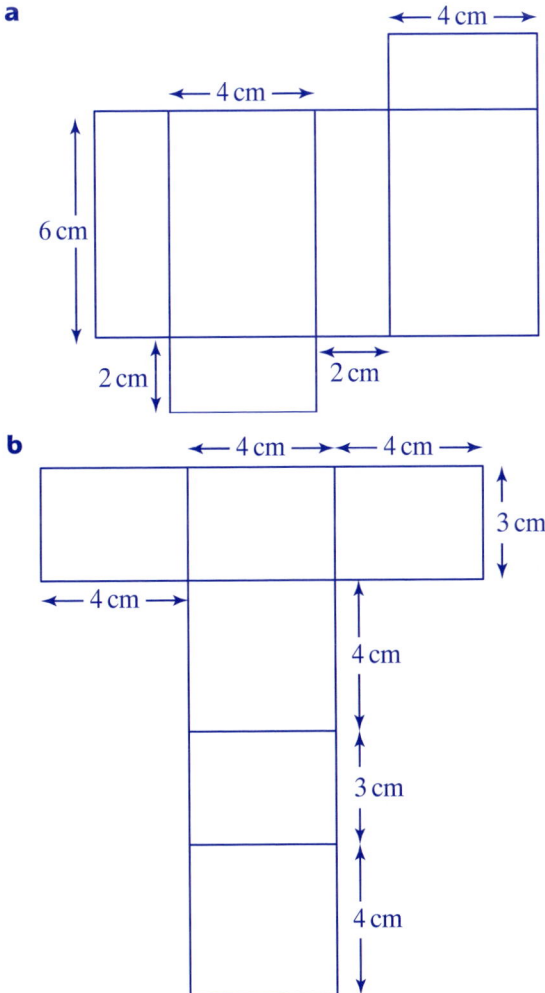

6 Get Real!

The isometric diagrams below are scale drawings of two workshops.

Sketch each workshop and give the dimensions on your sketch.

a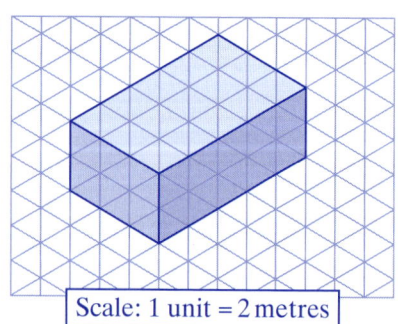

Scale: 1 unit = 2 metres

b

Scale: 1 unit = 2 metres

7 A cuboid is 12 cm long, 10 cm wide and 20 cm high.

a Draw an isometric diagram of this cuboid using 1 unit on the grid to represent 2 cm.

b Sketch a net of the cuboid.

8 A teacher asks students to draw an isometric drawing of a cube with sides of length 2 cm, including unseen edges as dotted lines.

Two of the attempts are shown below. What is wrong with each diagram?

a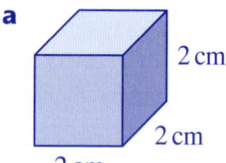

2 cm

2 cm

2 cm

b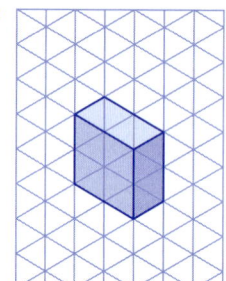

9 On an isometric grid, draw two **different** cuboids that have a volume of 48 cm³.

10 Get Real!

A manufacturer wants to make bars of soap, each with a volume of 216 cm³.

Draw an isometric diagram to show the possible dimensions of a bar if it is in the shape of:

a a cuboid   **b** a cube.

11 The T-shape shown in the diagram is made from cuboids.

    a Sketch the shape.

    b Calculate:

        i its volume

        ii its total surface area.

    c Write down the number of:

        i faces

        ii edges

        iii vertices.

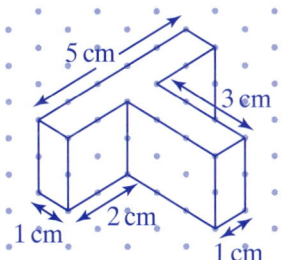

Homework 2

1 Draw a sketch of:

    a a square-based pyramid

    b a hemisphere

    c a pentagonal prism.

2 Sketch a prism whose cross-section is:

    a an isosceles triangle

    b a rhombus

    c an isosceles trapezium.

3 Each of the 3-D solid below is made from six centimetre-cubes.

    On a centimetre grid, draw the plan of each solid and its elevations from the directions marked F and S.

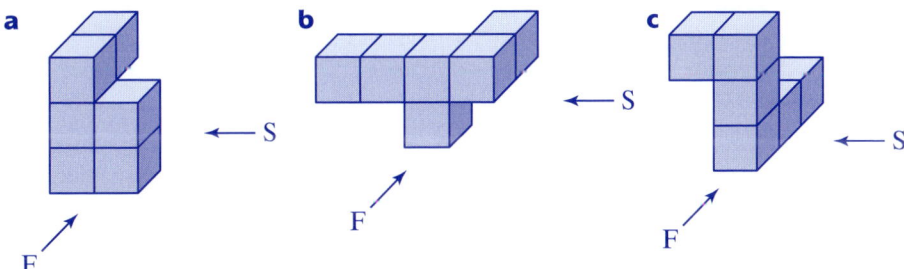

4 Ten centimetre-cubes are arranged into the solid shown in the diagram.

On a grid draw:

a the plan of this 3-D solid

b the front elevation as viewed from A

c the side elevation as viewed from B.

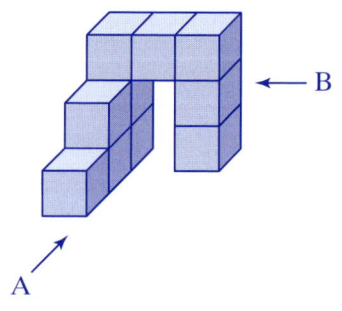

5 Get Real!

The diagram shows the dimensions of a can of tuna.

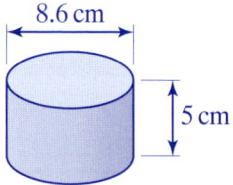

Draw an accurate plan and elevation. Leave out the hidden edges.

6 For each prism below, sketch the plan, front and side elevations from the directions shown by the arrows.

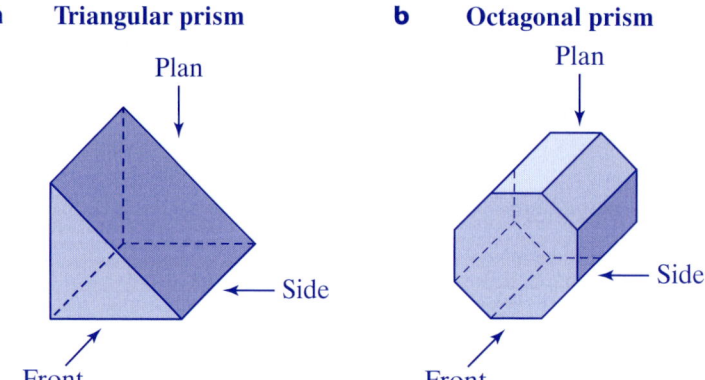

a   Triangular prism

Plan

Side

Front

b   Octagonal prism

Plan

Side

Front

7 Each solid below is a cuboid with part removed. The dimensions are given in centimetres.

For each object, draw full-size plan, front and side elevations on a centimetre grid.

a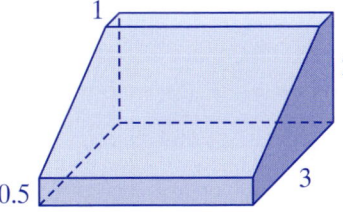

b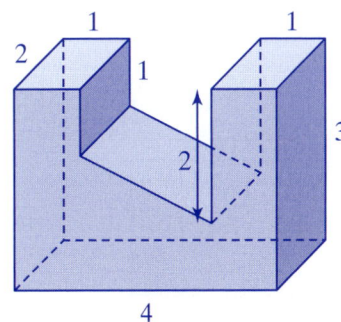

8 Tom has drawn a plan and a side elevation of the 3-D solid shown below.

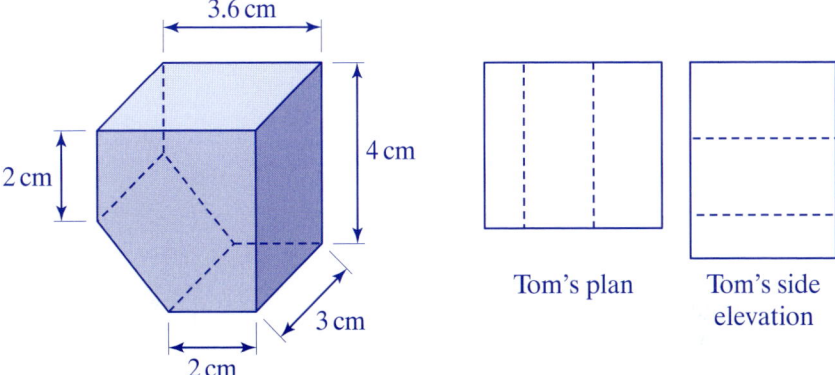

Tom's plan   Tom's side
                    elevation

a Describe what is wrong with Tom's diagrams.

b Draw an accurate plan and side elevation.

9 Plans and elevations of three prisms are shown below. Sketch each object.

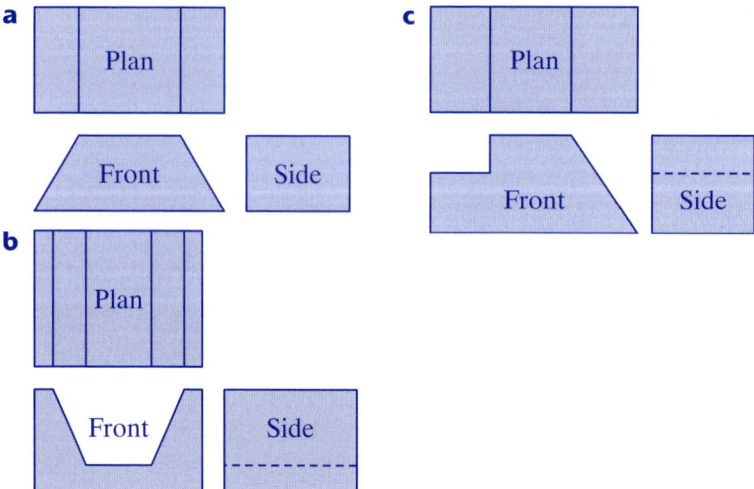

10 Get Real!

The diagram shows a toolbox.

a Sketch a plan of the toolbox.

b Sketch an elevation of the toolbox when it is viewed from the front, F.

c Sketch an elevation of the toolbox when it is viewed from the side, S.

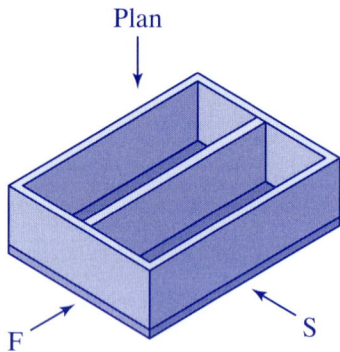

17 Algebraic proofs

Homework 1

1 m and n are both odd numbers.

m is greater than n.

Is $m - n$ always odd, always even or could it be either?

2 v and w are both even numbers.

Is vw always odd, always even or could it be either?

3 Is the value of $3n + 1$ always an even number?

Give a reason for your answer.

4 Is the value of $5n - 1$ always an even number?

Give a reason for your answer.

5 If q is an even number explain why $(q - 1)(q + 1)$ is an odd number.

6 P is a prime number.

Q is an even number.

State whether each of the following is:

i always even

ii always odd

iii could be either even or odd.

    a PQ

    b P(Q + 1)

7 The product of two consecutive numbers is divisible by 30.

Write down two numbers that satisfy this statement.

8 Dewi says that the sum of two prime numbers is always even.

Give a counter example to show that Dewi is wrong.

9 Sue says that the square root of a number is always smaller than the number itself.

Is Sue correct?

Give a reason for your answer.

10 Pam says that the cube of a number is always bigger than the number itself.

Is Pam correct?

Give a reason for your answer.

11 Prove that the sum of three consecutive numbers is always a multiple of three.

12 Prove that the sum of two consecutive odd numbers is always an even number.

13 Prove that the product of any two consecutive even numbers is always a multiple of 4.

14 Jan says that $5(a + b) = 5a + b$

Is Jan correct?

Give a reason for your answer.

15 Andrew says that the difference between two prime numbers is always an even number.

Is Andrew correct?

Give a reason for your answer.

16 Margaret says that if n is an odd number then $4n - 1$ is a prime number.

Is Margaret correct?

Give a reason for your answer.

17 Part of a number grid is shown below:

1	2	3	4	5	6	7
8	9	10	11	12	13	14
15	16	17	18	19	20	21
22	23	24	25	26	27	28
29	30	31	32	33	34	35
36	37	38	39	40	41	42
43	44	45	46	47	48	49

The shaded cross is called C_{10} because it has 10 in the centre.

a This is C_n

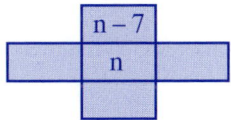

Copy and fill in the empty boxes on C_n

b Neeta notices that $3 + 11 + 17 + 9 = 40$

and that $4 \times 10 = 40$

Show, using algebra, that the sum of the arms of any cross is equal to four times the number in the centre of the cross.

Using and applying mathematics coursework task

As part of your GCSE course you are required to submit two coursework tasks covering:

- Using and applying mathematics
- Handling data

Each coursework task is worth 10% of the total marks for your GCSE so it is important that you spend time on each piece of coursework and try to get as high a mark as possible.

This chapter explains how to get the best marks in the Using and applying mathematics coursework task. The Handling data coursework task is discussed in Homework Book 1.

Marks are awarded for the coursework task under three headings, called **strands**. Each strand is marked out of 8 marks. At the foundation tier, you should aim to get between 3 and 6 marks.

Strand 1   Making and monitoring decisions to solve problems

This strand is about deciding what needs to be done and then doing it. You need to select an appropriate approach, find information and introduce questions of your own that develop the task further.

Mark	Making and monitoring decisions
3	You must consider the given task and obtain the necessary information to solve it.
4	Break down the given task and solve the task in an orderly manner.
5	Introduce your own *relevant* question(s) to develop the task beyond the given task.
6	Develop and follow through alternative approaches making use of more demanding mathematics.
*7	Analyse and give reasons for your alternative approaches, making use of a number of mathematical variables or features.
*8	Explore independently and extensively a range of appropriate mathematical approaches to the task.

At the foundation tier, you should aim to get between 3 and 6 marks.
* The content for marks of 7 and 8 is included here for information.

Strand 2   Communicating mathematically

This strand is about communicating what you are doing using words, tables, diagrams, graphs and symbols (algebra). Your chosen presentation should be used to identify patterns and provide generalisations.

Mark	Communicating mathematically
3	Collect together your information using tables, diagrams, graphs or symbols (algebra).
4	Use *appropriate* presentation along with linking commentary and interpretation.
5	Make use of algebra in forming generalisations and using substitution to check them.
6	Use algebra accurately to support the work and provide justifications.
*7	Use higher level algebra accurately in presenting a reasoned argument for your work.
*8	Use higher level algebra concisely and efficiently in presenting a reasoned argument for your work.

At the foundation tier, you should aim to get between 3 and 6 marks.
* The content for marks of 7 and 8 is included here for information.

Strand 3   Developing skills of mathematical reasoning

This strand requires you to search for patterns and provide generalisations for your task. Generalisations should then be tested on new data, justified and explained.

Mark	Developing skills of mathematical reasoning
3	Use the information collected to make a generalisation that is true for all the results.
4	Check the generalisation by testing a further example involving new data.
5	Provide a justification for the generalisation; the justification can be algebraic, graphical or diagrammatic.
6	Draw together the generalisations for the extended task and provide a justification for these.
*7	Provide an overarching justification that coordinates a number of mathematical variables or features.
*8	Provide a mathematically rigorous justification, argument or proof that includes the conditions for its validity.

At the foundation tier, you should aim to get between 3 and 6 marks.
* The content for marks of 7 and 8 is included here for information.

Explore 1   Number grid

Look at this number grid:

1	2	3	4	5	6	7	8	9	10
11	12	13	14	15	16	17	18	19	20
21	22	23	24	25	26	27	28	29	30
31	32	33	34	35	36	37	38	39	40
41	42	43	44	45	46	47	48	49	50
51	52	53	55	55	56	57	58	59	60
61	62	63	64	65	66	67	68	69	70
71	72	73	74	75	76	77	78	79	80
81	82	83	84	85	86	87	88	89	90
91	92	93	94	95	96	97	98	99	100

⊚ A box is drawn around four numbers

⊚ Find the product of the top left number and the bottom right number in this box

⊚ Do the same with the top right and bottom left numbers

⊚ Calculate the difference between these products

Investigate further

This task is an AQA set task so it can be submitted for marking by AQA or marked by the centre.

The words 'Investigate further' mean that you should develop the task beyond its original scope.

There are no right or wrong ways in which the task can be developed. Think of different ways in which you might extend the task.

Some tips

You may wish to consider the following:

⊚ Think about how you might extend the task. In what ways might the task be developed? Try out some of these ways and remember to comment on your findings.

⊚ Remember to include tables (or graphs) to illustrate your findings. Use your tables (or graphs) to write down any patterns that you notice. Remember to show all your working.

⊚ Think about your findings and try to explain them. Are you sure that the results are correct? How do you know? Have you tried all of the possible variations?

⊚ Use algebra to develop the task further. Can you write the product in algebraic terms? Can you write the difference in algebraic terms? Make sure you explain what you are doing.

Explore 2 Trays

A shopkeeper asks a company to make some trays.

A net of a tray made from a piece of card measuring 18 cm by 18 cm is shown below:

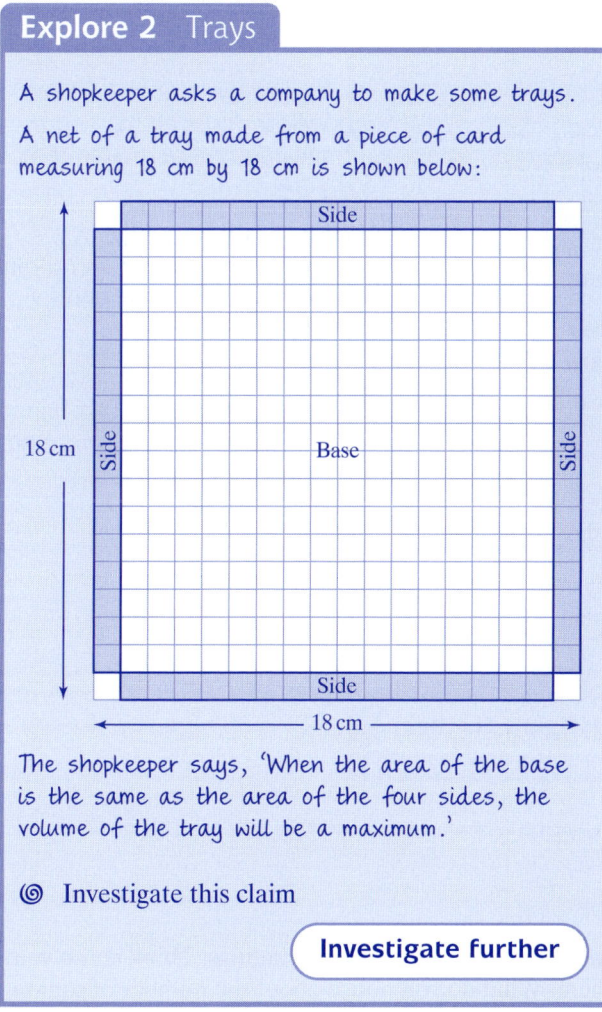

The shopkeeper says, 'When the area of the base is the same as the area of the four sides, the volume of the tray will be a maximum.'

◎ Investigate this claim

(**Investigate further**)

This task is an AQA set task so it can be submitted for marking by AQA or marked by the centre.

The words 'Investigate further' mean that you should develop the task beyond its original scope.

There are no right or wrong ways in which the task can be developed. Think of different ways in which you might extend the task.

Some tips

You may wish to consider the following:

◎ Use squared paper to create the net of the tray, then use your net to make the box. What is the area of the base? What is the area of the sides? What is the volume of the tray? What do you notice?

◎ Think about how you might extend the task. In what ways might the task be developed?

◎ Remember to include tables or graphs to illustrate your findings. Use your tables or graphs to write down any patterns that you notice. Remember to show all your working.

◎ Use algebra to develop the task further. Can you write the areas in algebraic terms? Can you write the volumes in algebraic terms? What do you notice? Make sure you explain what you are doing.

Explore 3   Tiles

Angela is making patterns from square tiles.

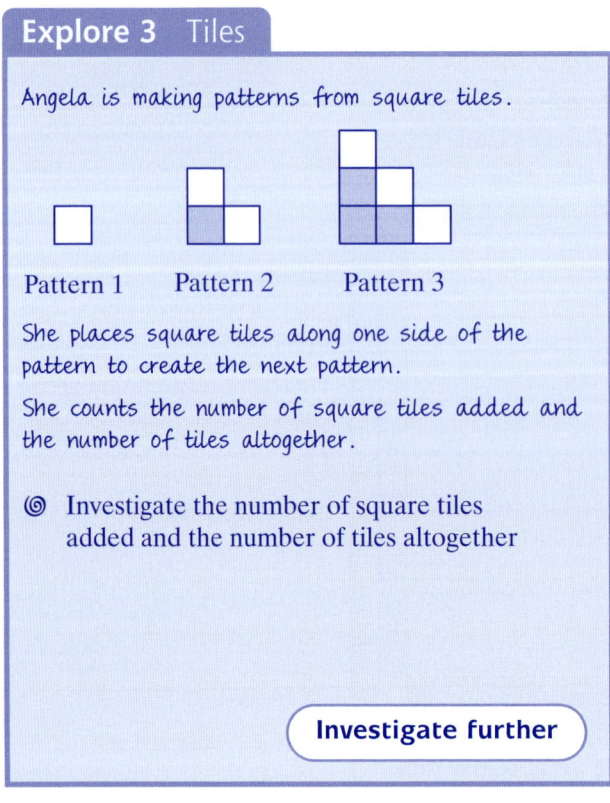

Pattern 1   Pattern 2   Pattern 3

She places square tiles along one side of the pattern to create the next pattern.

She counts the number of square tiles added and the number of tiles altogether.

◎   Investigate the number of square tiles added and the number of tiles altogether

This task is not an AQA set task so must be marked by the centre and moderated by AQA.

The words 'Investigate further' mean that you should develop the task beyond its original scope.

There are no right or wrong ways in which the task can be developed. Think of different ways in which you might extend the task.

Investigate further

Some tips

You may wish to consider the following:

◎   Try out some further patterns and make a note of your findings. Think about your findings and try to explain them. What do you notice about the number of square tiles added and the number of tiles altogether?

◎   Remember to include tables (or graphs) to illustrate your findings. Use your tables (or graphs) to write down any patterns that you notice. Remember to show all your working.

◎   Think about how you might extend the task. In what ways might the task be developed? Try out some of these ways and remember to comment on your findings.

◎   Use algebra to develop the task further. Can you write down the nth term for the number of square tiles added and the nth term for the number of tiles altogether? Make sure you explain what you are doing.